Freeze-Drying Biological
Specimens:
A Laboratory Manual

Frontispiece: Figure 1. Pallas Cat, *Felis manul.*

Freeze-Drying Biological Specimens: A Laboratory Manual

By Rolland O. Hower
Office of Exhibits Central
Smithsonian Institution
Washington, D.C.

Introduction by
R. H. Harris
British Museum of Natural History
London, England

Smithsonian Institution Press   City of Washington   1979

Library of Congress Cataloging in Publication Data

Hower, Rolland O.
    Freeze-Drying Biological Specimens: A Laboratory Manual.

    Bibliography: p.
    Includes indexes.
    1. Freeze-drying. 2. Biological specimens—
Collection and preservation. I. Title.
QH324.9.C7H68   579′.2   78-10750
ISBN 0-87474-532-2

Cover: Freeze-dried specimen of Maxwell screech owl *(Otus maxwellii)*

Photograph by James A. Mahoney, Chief, Office of Exhibits Central, Smithsonian Institution.

All other photographs by the author.

Typeface: Times Roman
Paper: Paloma 70 lb. text, coated matte
Designed by Natalie Babson
Printed by Eastern Press, Inc.

First printing, 1979
Second printing, 1981

Fondly dedicated to the memory of
John E. Anglim
who shared the dream,
supported the research, encouraged the author,
and was most of all a friend.

CONTENTS

Figures

Tables

FOREWORD

Conservation, in all the senses that the word implies, is perhaps the most essential function of a museum. The objects in a museum's custody are testimonials of past developments in nature and of the accomplishments of humanity. Hence, it is not only fitting but incumbent on any museum that has the resources to do so to seek ways to develop methods which might lead to more efficient preservation of the heritage in its custody, whether it be the product of human activity or of nature.

In some cases, the development of proper conservation methods is integral to the development of exhibition techniques. This is certainly the case of taxidermy which, for many years, provided the only method by which the lifelike appearance of the animal world could be three dimensionally shown in a museum. Yet this was but an imperfect solution, for only the appearance was maintained: the exterior shell—the skin of the animal—suitably stretched over an appropriate support. Biological reality was lost in the process, and in those cases where it was deemed desirable to fully maintain it, the specimen was kept submerged in a preserving liquid in a glass container. Yet there was another methodology that could present the lifelike appearance of the specimen and, at the same time, preserve its most essential parts—freeze-dry "taxidermy," a technique whose potential was recognized by the Smithsonian Institution and a technique in which the author of this volume, Rolland O. Hower, has become one of the foremost practitioners. In these pages he presents the results of decades of experience and collaboration with colleagues having similar interests, and gives the guidance that can enable others to begin to follow in his footsteps.

Time will tell to what extent this technology can be developed further and what its ultimate impact will be on the ability of future scholars to more fruitfully study the animal evidence collected during this generation. No doubt, however, being able to study all the parts in their relative position, unaltered by any major infection of chemical compounds, can only expand the usefulness of these specimens for future research.

This book marks a significant stage in the development of this new technology. It is hoped that there will emerge a later edition benefitting from the added experience gained by those who use and perfect further the techniques described in these pages.

February 6, 1978   Paul N. Perrot
                          Assistant Secretary for Museum Programs,
                          Smithsonian Institution

PREFACE:

A BRIEF SURVEY OF FREEZE-DRY RESEARCH

At the Smithsonian Institution, the freeze-dry techniques used to preserve biological specimens had their beginnings in the late 1950s. The total experience since then is best measured in terms of the more than four thousand specimens that have been preserved by these techniques and that now provide a vast resource of data on all types of biological tissues.

This book contains the most pertinent examples of the freeze-dry technique, and summarizes present knowledge of the methodology, including new, hitherto unpublished, information.

It is well to establish at once what is meant by the term "freeze-drying." Freeze-drying is a technique by which biological specimens are frozen, then dehydrated by sublimation, with complete retention of physical detail and form. Sublimation is the process by which the water in cells passes from a frozen to a gaseous state without going through a liquid state. No chemical agents are required. Unlike tissue dried from a nonfrozen state, freeze-dried biological tissue does not shrink or otherwise become distorted.

Biological entities ranging from amoebas to alligators have been preserved by freeze-drying. Except for some smooth-skinned animals such as certain amphibians and fishes, physical distortion is so negligible that it is easily corrected.

Students and researchers involved with freeze-drying gross pathology specimens or massive tissues of whole biological specimens are often perplexed and confused by the profusion of data, much of which seems contradictory. Literature describing shell or thin-layer freezing and the subsequent lyophilization of pharmaceuticals, foods, and other materials indicates that the latent heat of sublimation is important in establishing the surface temperature of specimens. This is valid for minimal masses—but not for the more massive tissues discussed here. Thus, food processors perfecting sophisticated preservation methods have contributed to certain misconceptions. From the point of view of workers dealing primarily with biological specimens, gross histological integrity—structure retention—is the primary consideration, and the factors dealing with the economics of food production are of no consequence.

Since the terms "vacuum dehydration" and "freeze-drying" occur throughout, it may be useful to explain them here. Vacuum dehydration occurs when frozen tissue is placed in an evacuated tank or desiccator over a suitable desiccant. (Concentrated sulfuric acid was originally used; but silica gel, anhydrous calcium sulfate, or phosphorous pentoxide are now used.) The desiccator is kept at room temperature, and the tissue must be permeable. In its simplest terms, freeze-drying occurs when a tank is placed in a deep freezer at temperatures below $-10°$ C and is connected to a vacuum source similar to that of the vacuum-dehydration apparatus. But sublimation in this case is carried out at a controlled freezing temperature. This technique is thus used for tissue that is not readily permeable.

Scanning electron microscopy (SEM), a relatively recent tool of science, offers an exciting new area for freeze-dry research. The scanning electron microscope was originally a development of the scanning electron microprobe. The first noteworthy SEM photographs of fibers and paper materials were produced in 1958. By 1965 scanning electron microscopes were available commercially and were being used for a wide variety of materials. Initially, soft biological materials were chemically fixed and simply air dried, then coated for examination in the SEM. Drawbacks were that tissue structures collapsed beyond a usable extent and smaller appendages and hairs were plastered against the bodies or tissue masses by surface tension during drying. It was only when freeze-dried and critical-point dried specimens could be produced under controlled conditions that the scanning electron microscope could be effectively applied to study 'soft biological tissues.

* * *

In important ways many persons have contributed to this volume. The person to whom I am most indebted is the late John E. Anglim, who was the Smithsonian's first Exhibits Director. I am also greatly indebted to R. H. Harris, Senior Experimental Officer at the British Museum of Natural History, who wrote the introduction and assisted in the preparation of the references. For many years he has been a friend and colleague. Over the past decade, our cooperative and, at the same time, independent efforts have made it possible for us to conduct many frutiful experiments and develop effective procedures. Through the years I have received invaluable help from H. T. Meryman, formerly of the Bio-physics Division of the Navy Medical Research Institute, who pioneered many freeze-dry techniques now in use. Professor Charles Howard, formerly of Catholic University of America and now the National Bureau of Standards, has assisted me greatly in many areas of research. Robert M. Organ, Chief of the Smithsonian Conservation and Analytical Laboratory, has offered constructive comments during the writing of this book. Dr. Norbert S. Baer of the Conservation Center of the Institute of Fine Arts, New York University, reviewed the manuscript of this book and contributed greatly to its present form. A. Gilbert Wright and Constance Minkin, Exhibits Editors with the Smithsonian's Office of Exhibits Central, have helped in providing a more readable text. Finally, I am indebted to Joyce Simon, Joyce Swartz, and Karen Hummer, who so carefully and speedily typed and retyped the manuscript.

Washington D.C.   Rolland O. Hower
January 1978

INTRODUCTION

by R. H. Harris

Although the sublimation of ice had been known as early as the middle of the 18th century and had been described by William Hyde Wollaston in 1813 in a communication to the Royal Society in London, the first authentic attempt to freeze dry tissue appears to have been made by Richard Altmann, who, at Leipzig in 1890, described a method of freeze-drying tissues to be used for histological examination.

The technique was, in fact, a form of vacuum dehydration in which the tissue, frozen at $-15°$ C, was placed in a desiccator over concentrated sulfuric acid. It was then evacuated by a pump with capability of at least 200 μm Hg. This is assumed because in later work, in which varying pressures were tried with vacuum dehydration, this pressure was shown to be the best; however, Altmann makes no mention of this.

Mann (1902) mentioned the technique in his book and attempted to repeat the experiment, but mechanical difficulties precluded his doing so.

Bayliss (1915) also experimented with the Altmann technique by dehydrating tissue in the same way. His intention was to study the replacement of water sublimed away in the dehydration process in the unfixed state, but he had much difficulty in his work owing to disintegration of the biological tissues. Romeis (1932) mentioned the method briefly in his book and apparently attempted to employ the technique, but was unsuccessful. In 1932, Gersh was the first to repeat the Altmann technique successfully. Previously Mann and Romeis had attempted and then abandoned the work.

Working in the laboratories of the University of Chicago (Department of Anatomy), Gersh overcame the various difficulties and procured good histological information from a number of tissues. Between 1933 and 1941, Bensley, Hoerr, and Gersh carried out a series of studies on cell morphology, mitochondria, chemical basis of cell organization, and the structure of the resting nucleus. They found that "fixation was more satisfactory in animals on total starvation for several days."

In 1935, Goodspeed and Uber adopted the Altmann technique for preliminary fixation for cytological studies and for microincineration work. Scott used the method for similar purposes in the preceding year. Hoerr (1936) wrote a very comprehensive paper in which all the recent developments up to the time were reviewed. The main findings were that far better results were obtained by rapid freezing, chilling with isopentane cooled with liquid nitrogen. Dehydration of tissue below $-30°$ C permits tissue to warm up from freezing to the dehydration temperature, avoiding the need for paraffin wax embedment.

Simpson (1941) published a paper concerned with determining the factors that might have affected the end result of Gersh's work in 1932, when he introduced the modern form of the Altmann technique. He came to the following conclusions: "The nature of the coarsest ice crystal reticulation is due to freezing and not to treatment afterwards." This was established by observing: "That the larger size of artifacts in tissue had a higher water content, and that there were not such artifacts in tissue prepared by a freeze thawing method. The rate of reticulation is governed by the rate of freezing, and the water content of the tissues, as well as changes in salt concentrations during freezing."

Simpson finally established that the preservation of cytological details in the outer layer of tissues frozen in isopentane cooled in liquid nitrogen surpasses in excellence chemically treated tissues. He further suggested that since tissues are subject to modification by heat, even after alcohol denaturization, embedding in collodion should be recommended as a routine procedure.

In 1948, Mercie, working in France, described a method whereby fungi were rapidly frozen and dried in a vacuum at $-10°$ C for three weeks. He kept his specimens in sealed jars since they were both brittle and hygroscopic.

Sjostrand (1951) started working with the Altmann technique and also carried out investigations using the modifications by Gersh. His paper was concerned with the structure of cells and the preservation of chemical components of cells. He used living cells as a reference standard, summarizing his work by stating, "Freeze-drying might be an adequate technique for the preservation of tissue cells for electron microscopy."

Sylven (1951) compared the now classic Altmann technique with standard procedures in histology. He states that the advantages are obvious, but that great care must be taken in the subsequent steps after drying. Davies (1954) prepared specimens by the Altmann method. He also experimented by cooling the desiccator to around $0°$ C. Due to the freezing of the sulfuric acid used as a desiccant, he tried phosphorus pentoxide and obtained excellent results with good color retention.

In 1956, Davies wrote another paper in which he described the successful preparation of animals and plants by drying from the frozen state. He made the first suggestion in this paper that small mammals might be dried in their entirety. Stadelmann (1959) repeated Mercie's work of 1948 and carried out carefully controlled methods for sealing the containers of dried fungi because of the hygroscopic nature of the material. Haskings (1960) described a method of preparing fungi by freezing at $-30°$ C overnight and then treating in an Edwards 3-PF freeze-dryer for three days, using a pressure of 5 to 10 μm Hg. Specimens were sprayed when finished with clear acrylic plastic.

Meryman (1960) gave the first description of entire specimen preparation of vertebrate animals together with work on plants, a very comprehensive paper which has formed the basis of all serious work in the field. This was followed a year later by the publication of technical details of the work described in the first paper. This enabled many workers to begin serious freeze drying of entire animals and plants for the first time. A good deal of confusion existed in the early 1960s as to exactly what was meant by "freeze-drying." There seemed to be several definitions, all different.

Hower, in 1962, in an interesting paper written as an information leaflet for the United States National Museum, in Washington, gave a general outline of what freeze-drying was all about. There were no technical details, just a simple explanation of how entire sepcimens might be dried without distortion and, in most cases, with virtually entire color retention. Blum and Woodring (1963), using a dry-ice condenser, carried out a successful series of experiments on insect larvae, followed later that year by another publication in which inexpensive apparatus was described and it was shown how arachnids (a particularly difficult group to preserve properly by any technique) might be dealt with.

Harris (1964), working in Great Britain, was the first to show in that country the use of freeze-drying in the whole biological range; following the work of (and after consultation with) Meryman, this paper dealt with both vacuum dehydration and freeze-drying throughout the plant and animal range.

Hower, in 1964, wrote an additional comprehensive account of developments at the U.S. National Museum. A description of a new apparatus for freeze-drying whole biological specimens was given in 1968 by Harris, and a paper on advances in freeze-dry preservation of biological specimens by Hower in 1970 followed many years of cooperation between Harris and Hower. Since then work has continued with much interest in the field of anatomical preparations and also in the preparation of soft tissue for sterioscan electron microscopy. This brief survey of many applications of the original Altmann technique of 1890 introduces the study of entire specimen freeze-drying.

Examples of rapidly permeable tissue are fungi, insect larval forms, etc. Relatively impermeable tissues are small mammals, birds, and flowering plants. Altmann's early work and that of his successors was vacuum dehydration; Meryman, Davies, Hower, and Harris are working with freeze-drying and it is this technique that now has the most prominent position in this field of preservation.

I

FUNDAMENTALS OF FREEZE-DRY

FREEZING BIOLOGICAL TISSUE

Of all the compounds in animal tissue, water is the most abundant. On the average, it constitutes 70 percent of the animal's total weight. Water is found in cellular and vascular spaces and, in small quantities, in protein and carbohydrates. Of the total liquid content, 20 percent is usually extracellular fluid (approximately 25 percent of this is plasma), and the remainder is interstitial fluid, mostly water. About 76 percent of muscle tissue is water. The Rowntree data (Harrow, 1951) relating to the biochemistry of man provides a comprehensive survey of the distribution of tissue water in most mammals.

Water in tissue is almost always found in combination with naturally occurring salts. For this reason, freezing should be as rapid as possible, since slow freezing invariably leads to concentration of salts and, therefore, a lower freezing point. As freezing occurs, water freezes out of solution by a process of ion diffusion and the solution becomes increasingly more salt-saturated. Salt concentration also reduces the vapor pressure, producing a longer drying cycle that may cause harmful shrinkage (owing to the unfrozen saturated solution in the tissue and the resulting surface tension).

Rapid freezing can be achieved by one of the freezing mixtures listed in Table 1, or by using a freezer chest with a temperture below $-25°$

TABLE 1.—FREEZING MIXTURES[1]

Primary substance		Secondary substance		Temperatures[2]			
				A	B	C	D
$NaC_2H_3O_2$ (cr)	85	H_2O	100	10.7	4.7		
NH_4Cl	30	H_2O	100	13.3	-5.1		
$NaNO_3$	75	H_2O	100	13.2	-5.3		
NaS_2O_3 (cr)	110	H_2O	100	10.7	-8.0		
KI	140	H_2O	100	10.8	-11.7		

TABLE 1.—FREEZING MIXTURES[1]

Primary substance		Secondary substance		Temperatures[2]			
				A	B	C	D
$CaCl_2$ (cr)	250	H_2O	100	10.8	−12.4		
NH_4NO_3	60	H_2O	100	13.6	−13.6		
K_2SO_4	10	snow	100	−1	−1.9		
$NaCO_3$ (cr)	20	snow	100	−1	−2.0		
KNO_3	13	snow	100	−1	−2.85		
$CaCl_2$	30	snow	100	−1	−10.9		
NH_4Cl	25	snow	100	−1	−15.4		
NH_4NO_3	45	snow	100	−1	−16.75		
$NaNO_3$	50	snow	100	−1	−17.75		
$NaCl$	33	snow	100	−1	−21.3		
$H_2SO_4 + H_2O$	1	snow	1.097	−1	−37.0	−37.0	.0
(66.1% H_2SO)	1	snow	1.26	−1	−36.0	−30.2	17.0
(66.1% H_2SO)	1	snow	1.38	−1	−35.0	−25.0	27.0
(66.1% H_2SO)	1	snow	2.52	−1	−30.0	−12.4	133.0
(66.1% H_2SO)	1	snow	4.32	−1	−25.0	−7.0	273.0
(66.1% H_2SO)	1	snow	7.92	−1	−20.0	−3.1	553.0
(66.1% H_2SO)	1	snow	13.08	−1	−16.0	−2.1	967.0
$CaCl_2 + 6H_2O$	1	snow	.35	0		0	52.1
$CaCl_2 + 6H_2O$	1	snow	.49	0		−19.7	49.5
$CaCl_2 + 6H_2O$	1	snow	.61	0		−39.0	40.3
$CaCl_2 + 6H_2O$	1	snow	.70	0		−54.9*	30.0
$CaCl_2 + 6H_2O$	1	snow	.81	0		−40.3	46.8
$CaCl_2 + 6H_2O$	1	snow	1.83	0		−21.5	88.5
$CaCl_2 + 6H_2O$	1	snow	2.46	0		−9.0	192.3
$CaCl_2 + 6H_20$	1	snow	4.92	0		−4.0	392.3
Alcohol at 4°C	77	snow	73	0	−30.0		
Alcohol at 4°C		CO_2	solid		−72.0		
Chloroform		CO_2	solid		−77.0		
Ether		CO_2	solid		−77.0		
Liquid SO_2		CO_2	solid		−82.0		
NH_4NO_3	1	H_2O	.76	20	5.0		
NH_4NO_3	1	H_2O	.94	20	−4.0		
NH_4NO_3	1	H_2O	.94	10	−4.0		
NH_4NO_3	1	H_2O	.94	5	−4.0		
NH_4NO_3	1	snow	.94	0	−4.0		
NH_4NO_3	1	H_2O	1.20	10	−14.0		
NH_4NO_3	1	snow	1.20	0	−14.0		
NH_4NO_3	1	H_2O	1.31	10	−17.5		
NH_4NO_3	1	snow	1.31	0	−17.5		
NH_4NO_3	1	H_2O	3.61	10	−8.0		
NH_4NO_3	1	snow	3.61	0	−8.0		

[1]Modified from Smithsonian Tables, 1959, Forsythe.
[2]C° throughout.
A = Before mixture.
B = Mixture.
C = When all snow is melted.
D = Heat absorbed (calories where A is given in grams).

C. Rapid freezing also creates smaller ice crystals, causing less tissue distortion. A temperature of $-30°$ C appears to be the optimum. Attempts to achieve rapid freezing with extremely low temperatures, such as would result from total immersion in liquid nitrogen, is almost certain to fracture specimens when the mass is greater than 1 gram. Also, as observed by Meryman (1960), when specimens are frozen at extremely low temperatures, the tendency for extracellular crystal formation is greatly reduced, and numerous smaller crystals begin to appear within the cells. This condition slows the drying process greatly.

THERMODYNAMICS OF WATER: ITS PROPERTIES IN VARIOUS STATES

The freeze-dry process ultimately depends upon the relationship of pressure and temperature to the molecular dynamics of water, discussed here without regard to the presence of inorganic salts. Due to hydrogen bonding, pure water is a liquid that maintains certain traces of crystallinity. The structure of ice is an OH_4 tetrahedron that is apparently distorted so that two of the hydrogen atoms are closer than the other two to the central oxygen atom. These four hydrogen atoms are located at the corners of the tetrahedron and are linked by O-H-O bonds. The bonding of the two distant hydrogen atoms is the hydrogen bond. The structure is described by Andrews and Kokes as being similar to that of quartz.

As ice melts, the average energy of the water molecules increases. This increase disturbs the orderly lattice through the rupture of some hydrogen bonds; the forces that have produced the tetrahedral structure of ice remain operative, however, as do traces of the tetrahedral arrangement. More specifically, melting the ice destroys the large crystal structure and forms numerous microcrystallites. In our example we must recognize that the constant association and dissociation of water molecules both structure and destroy these crystallites. Therefore, they exist only statistically.

As the temperature is increased, the average kinetic energy increases, and molecular motion becomes violent; the probability of molecules escaping from the liquid mass increases. As the boiling point is reached, maximum vaporization occurs.

The escape of molecules at high velocity requires a transfer of energy from the liquid mass; it is necessary to apply a specific quantity of energy to convert a given quantity of water to vapor. (At $100°C$, 540 calories are required to convert 1 gram of water [latent heat of vaporization] to the gaseous state).

The boiling point of water is $100°$ C at a pressure of 1 atmosphere (760 mm Hg). Conversely the vapor pressure of water may be described as 760 mm Hg at a temperature of $100°$ C.

When pressure and temperature are held constant, a state of equilibrium is established. The interface between the vapor and liquid is controlled by the surface tension produced by the geometric lattice of intermolecular bonds at the liquid surface.

The escape of molecules may be increased to the maximum by changing either of two conditions. As described previously, if the temperature is increased, a maximum number of water molecules will escape into the atmosphere. Or, if the pressure over the water is reduced, a similar escape is predictable. Sublimation will occur at temperatures below the freezing point of water and at suitable pressures.

VAPOR PRESSURE AND SUBLIMATION

If a closed chamber containing ice is maintained at a given temperature and evacuated with a vacuum pump, then valved off so that no external air can enter, molecules will escape from the ice within the chamber to form water vapor. This escape (sublimation) will continue as the free molecules collide with one another and with the walls of the chamber. Many of these free or escaped molecules striking the surface will reassociate with the ice crystals.

Eventually the rate at which molecules return to the ice will equal the rate at which they escape. The water vapor and ice are then said to be in a state of equilibrium. The pressure of the vapor at which this equilibrium occurs is referred to as the *equilibrium vapor pressure*.

If for example the temperature in the chamber is $-15°$ C, the pressure will reach 1.241 mm Hg. This is the equilibrium vapor pressure (consequently the vapor pressure) at $-15°$ C. It should be noted that the amount of heat required to convert 1 gram of ice to a vapor is 620 calories (latent heat of sublimation) which is equal to the combined latent heats of fusion and vaporization (80 and 540 calories respectively).

Although the chamber is refrigerated at temperatures below freezing, the refrigerated surface is, in fact, a heat source. The heat is transferred from the refrigerated chamber surface to the surface of the specimen where this energy is used for the sublimation of ice crystals.

The main difference between freeze-drying massive biological tissues, described in this text, and food and pharmaceutical applications is the temperature used for sublimation.

TABLE 2.—THERMODYNAMIC PROPERTIES OF WATER

| T°K[2] | T°C | Vapor pressure[1] | | | Specific volume[3] saturated vapor |
		ICT	Kelly	Hower	
273.15	0	4.570	4.56472	4.579	206.31
272.15	-1	4.217	4.20340	4.216	221.09
271.15	-2	3.880	3.86831	3.879	237.14
270.15	-3	3.568	3.55772	3.567	354.45
269.15	-4	3.280	3.27002	3.2779	273.22
268.15	-5	3.013	3.00367	3.01028	293.51

TABLE 2.—THERMODYNAMIC PROPERTIES OF WATER (cont.)

T°K[2]	T°C	Vapor pressure[1]			Specific volume[3] saturated vapor
		ICT	Kelly	Hower	
267.15	− 6	2.765	2.75725	2.76284	315.56
266.15	− 7	2.537	2.5294	2.5341	339.44
265.15	− 8	2.326	2.31886	2.32279	365.36
264.15	− 9	2.131	2.12443	2.1277	393.55
263.15	− 10	1.950	1.94501	1.94769	424.09
262.15	− 11	1.785	1.77953	1.78171	457.46
261.15	− 12	1.632	1.62701	1.62876	493.58
260.15	− 13	1.490	1.48653	1.4879	535.62
259.15	− 14	1.361	1.35724	1.3583	576.04
258.15	− 15	1.241	1.2383	1.239	623.05
257.15	− 16	1.132	1.12899	1.12956	674.31
256.15	− 17	1.031	1.02857	1.029	730.46
255.15	− 18	0.939	0.9364	0.9366	791.14
254.15	− 19	0.854	0.85186	0.8519	858.37
253.15	− 20	0.776	0.77347	0.77435	931.1
252.15	− 21	0.705	0.70339	0.7033	1012.15
251.15	− 22	0.64	0.6384	0.6383	1100.11
250.15	− 23	0.58	0.579	0.579	1196.17
249.15	− 24	0.526	0.524726	0.52445	1302.08
248.15	− 25	0.476	0.4751	0.4748	1418.44
247.15	− 26	0.43	0.4229	0.4295	1547.99
246.15	− 27	0.389	0.3886	0.3883	1689.19
245.15	− 28	0.351	0.351	0.351	1845.02
244.15	− 29	0.317	0.317	0.317	2016.13
243.15	− 30	0.2859	0.2857	0.2854	
233.15	− 40	0.0966	0.09679	0.0965	
223.15	− 50	0.02955	0.02976	0.0296	
213.15	− 60	0.00808	0.0082	0.0081	
203.15	− 70	0.00194	0.00199	0.00197	
193.15	− 80	0.00040	0.000418	0.000411	
183.15	− 90	0.000070	0.000074	0.000072	

[1]Given in mm Hg.   [2]Kelvin.   [3]Given in M[3] per kg.

A knowledge of vapor pressure at temperatures below the triple point[1] is essential. Table 2 contains vapor-pressure data for ice. Three sets of values are given. One set is from the International Critical Tables. The second set extends a few of the values to lower temperatures calculated by Kelly for comparison. The vapor pressure of ice is given by the relationship:

$$\log P = 1.207 + 3.857 \log T - 3.41 \times 10^{-3}T - 4.875 \times 10^{-8}T^2$$

$$- \frac{2461}{T}$$

T is absolute temperature (K°) and P is pressure given in μm Hg.

[1] Triple point—the point at which a solid, liquid, and vapor are in equilibrium (0° C at pressure of 4.579 mm Hg).

T is absolute temperatue (K°) and P is pressure given in μm Hg.

The third set of values is based on a semilogarithmic plot extrapolated from the ICT data by Hower with the help of Professor Charles C. Howard.

(a) $\log P = 10.430 - \dfrac{2668.5}{T}$ or an alternate form

(b) $P = 4.579 \exp\left[22.495\left(1 - \dfrac{273.15}{T}\right)\right]$

Form (a) has the advantage of a Kelly-type configuration for comparison purposes. Form (b) has the advantage of showing where the triple-point pressure (4.579) and temperature (-273.15) enter into the equation. The inverse of form (a) may be used to calculate the temperature associated with a given pressure:

$$T°K = \frac{2668.5}{10.430 - \text{Log } P}$$

THE FREEZE-DRY CONCEPT

Once frozen, biological tissue maintains mechanical rigidity, and as water is removed by sublimation most tissue can be dehydrated without apparent physical change.

To develop a definitive analysis of what actually occurs as biological tissue is freeze-dried, a broad generalized statement would be required. It would incorporate the following factors:

1. An exact statement of temperature variations along the sublimation boundary, with due consideration for the specific effects of the latent heat of sublimation.

2. The mean free path of water-vapor molecules within the complex and variable structure of biological tissue.

3. The nonlinear vapor-pressure variations of the biological fluids in various locations within the specimen.

4. The heat and mass transfer through the porous media of the dried tissue.

All are to be calculated at a given temperature and a specific pressure. Unfortunately, the complexity of this problem not only precludes an easy solution, but also prevents a satisfactory method of formulating the problem.

At this point we must be content with delineating the principles as separate elements, with the hope that they ultimately will lead to a general understanding of the complex happenings within biological tissue as it is freeze-dried.

We have already mentioned the molecular structures and kinetic energy levels during the change in physical state (fusion and sublimation of ice) and the mechanisms of biological freezing. The next phase is the transfer of water-vapor molecules from the ice within the specimen to the condensing surface of the system.

TRANSFER OF WATER MOLECULES
TO SPECIMEN SURFACE

Sublimation begins at the outer surface of the specimen and continues at the boundary between the frozen and the dried tissue. This boundary recedes toward the center of the specimen as drying proceeds. As water molecules continue to escape from the ice crystals on the sublimation boundary, they move about at high velocity, colliding with one another and with the structure of the surrounding dried tissue. (As they are buffeted from collision to collision, they are virtually independent of external forces.) There is a heavy concentration of water-vapor molecules escaping; consequently, at the sublimation boundary there are more collisions, which ricochet molecules along the line of least resistance toward the outer shell of the specimen. The force of these collisions intensifies the movement of water-vapor molecules through the dried tissues of the specimen and into the atmosphere beyond its outer shell.

The next step is to remove the water-vapor molecules from the vicinity of the specimen. The most effective mehtod of continuously removing water vapor from the specimen chamber is to create a lower vapor pressure elsewhere. This is generally accomplished by establishing a colder surface (condenser) nearby. Water vapor will diffuse to the colder surface and recondense to form new ice crystals.

A refrigerated condenser serves as a very efficient vapor pump. When the temperature of a specimen chamber is $-15°$ C, the vapor pressure of the ice within the chamber is 1.241 mm Hg. If the temperature of the condensing chamber surface is $-40°$ C, the vapor pressure at that surface is 96 μm Hg, creating a vapor-pressure ratio of 13 to 1. With a pressure ratio this great, the only factor limiting the effectiveness of the condenser is the extent of its surface area.

Specimen and condensing-surface temperatures are establihshed to provide a working mean-vapor-pressure relationship. For example, in some scanning electron microscopy it is necessary to dry specimens at $-40°$ C or below to maintain histological integrity, thus requiring a substantially lower condenser temperature. It is recommended that the pressure ratios be at least 2 to 1.

At atmospheric pressure, the mean free path (average distance a molecule can travel before colliding with another molecule) is approximately .005 (5×10^{-3} mm) at a pressure of $.2\mu$ Hg, the mean free path is approximately 20 mm.

The transfer of water-vapor molecules from an ice crystal within the specimen to the condenser is greatly accelerated by extending the mean free path, easily accomplished by reducing the pressure with a vacuum pump.

Relationships of the mean free path to various pressures are given in table 3.

TABLE 3.—PRESSURE AND MEAN FREE PATH

Pressure	Mean free path
10 mm Hg	0.0034 mm
1 mm Hg	0.034 mm
100 μm Hg	0.34 mm
10 μm Hg	3.4 mm
1 μm Hg	34.0 mm

II
ENGINEERING SPECIFICATIONS

BASIC REQUIREMENTS OF A FREEZE-DRY SYSTEM

The apparatus used to preserve materials outlined in this text generally consists of a specimen chamber and a low-temperature water-vapor condenser joined by a vapor-conducting tube. A vacuum pump is connected to the condenser, and both the condenser and the specimen chamber are maintained at approximately the same low-pressure level. The chamber and the condenser are independently refrigerated to their appropriate temperatures.

Figure 2, an elementary schematic, shows the two basic systems dealing with pressure and temperature.

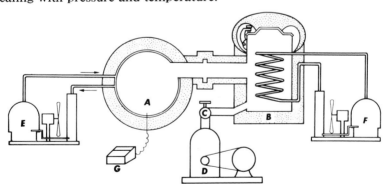

Figure 2. Elementary schematic of a freeze-dry system. *A*, specimen chamber; *B*, condenser; *C*, vacuum shutoff valve; *D*, vacuum pump; *E-F*, refrigeration compressors; *G*, Pirani gauge.

The Vacuum Pump

The vacuum pump, a vital part of any freeze-dry system, must be selected with great care. There are many suitable pumps available.

When a refrigerated condenser is used as a vapor trap, the primary function of the vacuum pump is to remove mechanically, noncondensable vapors and air from the system. Described in terms of displacement, expressed in liters per second or cubic feet per minute, a pump's capacity is the capability of a pump to remove a specific volume of gas from an evacuable space in a given amount of time (see tables 4 and 5).

The ultimate vacuum for a mechanical vacuum pump is the lowest pressure limit of the operating range of the pump in a closed system. In mechanical pumps the lowest pressure is limited by leakage through the discharge port back into the pump.

TABLE 4.—VACUUMS

Terminology	Pressure range
Ultrahigh vacuum	pressures below 1×10^{-6} mm Hg or .001 μm Hg
High vacuum	1×10^{-3} — 1×10^{-6} mm Hg
Fine vacuum	1 mm Hg 1×10^{-3} or 1μm Hg
Rough vacuum	760 mm Hg to 1 mm Hg
1 mm Hg	1000 m
	.00132 atmosphere
	.01934 psi
	.0393 in Hg
1 μm	.001 mm Hg
	1×10^{-3} mm Hg
Atmospheric pressure	760 mm Hg
	14.7 psi
	29.921 in Hg

TABLE 5.—CONVERSION MULTIPLIERS[1]

From	To	Factor
liters per second	cubic feet per minute	02.1200
liters per minute	cubic feet per minute	00.0353
liters per second	liters per minute	60.0000
cubic centimeters per second	liters per second	00.0010
cubic feet per minute	liters per minute	28.3200

[1]Liters per second $\times$ the factor $=$ cubic feet per minute.

Rotary Oil Pump

This type of vacuum pump (figure 3) consists of a cylindrical steel casing and a steel cylinder that rotates eccentrically within it. Mounted in the rotating cylinder, two spring-loaded vanes ride tightly against the inner walls of the steel casing. At the same time that air enters the intake port, it is carried to and expressed from the exhaust valve into the atmosphere.

As a thin film seal between the rotating and stationary parts, the oil in the pump serves as a lubricant, and prevents air from leaking back to the intake port.

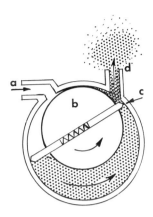

Figure 3. Rotary Oil Pump. *a,* intake; *b,* eccentric cylinder; *c,* vanes; *d,* exhaust.

With many variations, such as a rotary piston instead of a vane system, this type of vacuum pump is commonly found in laboratories throughout the world.

Rotary Piston Pump

The operating mechanism of the rotary piston vacuum pump is similar to the rotary oil pump, except that the piston type does not have vanes and the oil pump uses a sliding valve system. The rotary piston pump has three basic moving parts:
1. the shaft, upon which is keyed an eccentric cam;
2. the piston, with an integral hollow inlet extension; and
3. the slide pin (figure 4).

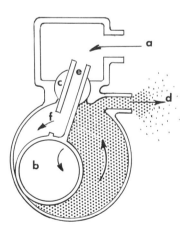

Figure 4. Rotary Piston Pump. *a,* intake; *b,* rotary piston; *c,* slide pin; *d,* exhaust; *e,* hollow inlet; *f,* inlet port.

As the shaft-and-cam assembly rotates, the hollow inlet extension passes freely within the slide pin in an up-and-down motion to open and close the inlet port. This rotating function serves as both an intake and an exhaust system.

Gas Ballasting

Sometimes called "vented exhaust," the gas ballast is essential in either type of vacuum pump if used for freeze-drying. Gas ballasting is no more than a technique for purging condensable or soluble vapor from the pump.

The condensation of vapors in the pump oil results from the high-compression characteristic of rotary pumps. Such condensed vapors in the oil reduce the efficiency of the pump; if water is allowed to accumulate, damage to the pump will occur. There are several sophisticated techniques for removing such vapors from the pump oil, but the gas ballast is the most effective and convenient.

The apparatus for gas ballasting is incorporated in the body of the pump and usually controlled with a small valve.

Air is admitted to the pump chamber after the vane or rotary cylinger has passed the intake port (figure 5). This air ballast lowers the effective compression ratio of the pump and thereby avoids the condensation of the water vapor before it is exhausted into the atmosphere.

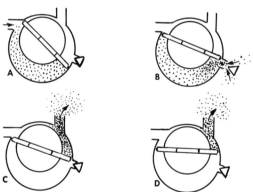

Figure 5. Schematic of Gas Ballast. *A*, preballast side of cycle; *B*, vanes block chamber to pump as ballast valve opens to admit air to exhaust side of pump cycle; *C*, vapor and air mixture released; *D*, exhaust cycle completed.

The reduction in compression ratio decreases the efficiency of the pump, and, therefore, the ultimate vacuum; for freeze-dry applications, however, the decreased efficiency is of no consequence.

Gas-ballasted pumps generally are rated with ultimate pressures from 1 to 10 μm Hg.

In pumps used for freeze-drying, oil should be changed after every 1000 hours of operation.

Oil Diffusion Pump

"Diffusion pump" is a specific term applied to vapor-stream pumps. These pumps are rarely used in freeze-drying gross biological tissues but are worth mentioning here for their value in preparing cell materials and microscopic biological specimens for study with a scanning electron microscope. This apparatus is described on pages 53 and 54.

An oil diffusion pump incorporates a high-velocity vapor stream to provide momentum to the pumped gas—in contrast to the mechancial methods of pumps described on preceeding pages.

The diffusion pump operates at low pressure; for this reason, it is used in conjunction with a mechanical roughing pump. Because of the geometric design of the nozzle and chamber of the diffusion pump, a stream of vapor is directed into a low-pressure chamber where it expands and converts pressure energy to velocity energy. The pumped gas enters the high-velocity vapor stream by molecular diffusion, resulting in the transfer of momentum to the pumped gas. The combined mass enters the diffuser, where velocity energy is converted back to pressure energy. The pumped gas is thus compressed.

Oil diffusion pumps—electrically heated and water-cooled—depend upon the vaporization of oils or esters to provide the vapor stream which directs the air molecules into the roughing pump; the oil vapor is then condensed and returned to the boiler.

Calculating Pump Capacity

Pump displacement or system speed must be calculated according to the volume of the system, the ultimate pressure, and the time required to pump the system to this pressure. See the pump-down chart (figure 6). Here are two examples:

(1) To evacuate a 15 cu ft system to a pressure of 100 μm Hg in 5 min., note that at 100 μm Hg the factor is 10.9. Multiply 10.9 by the volume (15 cu ft); 163.5 cu ft are to be pumped. To determine the pump displacement, divide this total by the time allowed for the evacuation (5 min).

$$10.9 \times 15 = 163.5 \text{ cu ft}$$

$$\frac{163.5}{5} = 32.7 \text{ cu ft/min or } 15.43 \text{ liters/sec}$$

(2) To determine how long it will take for a pump with a displacement of 27 cu ft per min to evacuate a system with a volume of 160 cu ft to a pressure of 100 μm Hg (at 100 μm Hg the factor is 10.9), calculate as follows:

$$10.9 \times 160 = 1744.0$$

$$\frac{1744.0}{27} = 64.59 \text{ min}$$

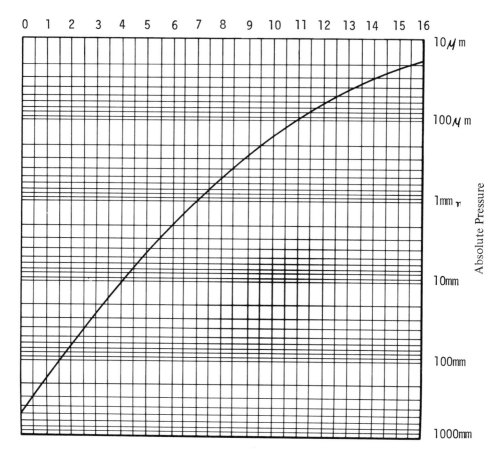

Figure 6. Vacuum Pump-Down Factors.

Under certain conditions (when very little water vapor is involved), the refrigerated condenser can be eliminated and water vapor can be removed directly through the vacuum pump. To know whether the condenser is needed, determine the approximate quantity of water vapor released each minute by the specimens in the chamber.

If, for example, the chamber contains three squirrels, two flickers, a toad, and a garter snake (at various drying stages and all of average weight), the average daily release of water vapor is approximately 12 g.

If 150 μm Hg of pressure were maintained within the chamber, it would be necessary to pump 200 cu ft of water vapor (see Table 1) for each gram of water removed from the system, or 2400 cu ft per day (1.66 cu ft/min). If the vacuum pump cannot handle this volume of vapor, the pressure inside the specimen chamber will slowly rise; as it rises, the volume occupied by a unit mass of water vapor decreases. At a pressure of 300 μm Hg, 2 g of water will occupy the same volume as would 1 g at half that pressure. To avoid increased operating pressure in the chamber, a refrigerated condenser should be used.

The differential pressure between the water vapor within the specimen and the pressure within the chamber provides the conditions for the movement of water vapor from the specimen. An increase in pressure within the chamber causes a decrease in efficiency.

If the specimen temperature is $-20°$ C and the chamber pressure is increased by 300 μm Hg, the pressure differential is then 380 μm Hg— about 40 percent less. This suggests the relationship whereby a vacuum limits the number of specimens that can be processed in a reasonable span of time.

Sealing Oil

Oil is generally used to seal and lubricate mechanical vacuum pumps. (Other liquids—such as tricresyl phosphate, lindol, or cellube—are used under special conditions.) To obtain the ultimate vacuum with mechanical pumps in freeze-dry systems, a suitable grade of water-free oil should be used.

Vacuum-Line Dimensions

The size of vacuum lines affects the efficiency of a freeze-dry system because a pump is effective only if the vacuum lines are large enough to handle the vapor transfer.

Conductance (the volume of flow of vapor through a tube) is a well-investigated function, and there are a variety of ways to calculate the vacuum-line dimensions required to permit adequate conductance.

The flow of fluids (or vapors) within a vacuum system is usually either molecular or viscous. Molecular flow occurs at pressures below 1 μm Hg; we are concerned only with viscous flow.

Restriction of tubing to the viscous flow of vapors is significant only when the volumes to be moved are very large or the pressure differential is small. The law that expresed the viscous flow of fluids through a tube was deduced by French physicist Jean L. M. Poiseuille. This law establishes relationships between the coefficient of viscosity (viscosity poises/dyne sec/cm²), the volume of the fluid flowing through the whole section of the tube in unit time (seconds, as applied here), the pressure differential at each end of the tube, the radius, and the length of the tube.

Q = flow rate (cm³/sec) d = diameter (cm)

P = pressure (dyne/cm²) L = length (cm)

$$Q = \frac{d^4}{256\ L}\ P$$

The volume of vapor transferred through a tube in a specific period of time is proportional to the fourth power of the tube radius. This means that a small increase in tube diameter will produce a considerable increase in vapor conductance.

To utilize the formula, it is necessary to know the coefficient of viscosity, which for air is approximately 1.7×10^{-4} poises at a pressure of 1 dyne/cm^2 = 7.5×10^{-4} mm Hg 1.

To calculate the conductance of vapor through a tube that is 193 cm long, 7.62 cm in diameter, and with a pressure differential between the specimen and condensing surface of 2000 μm Hg, the values are:

$$V = cm^3/sec$$

$$P = 2666.7 \text{ dynes/cm}^2 \ (1000 \ \mu\text{m Hg mean pressure})$$

$$d = 7.62 \text{ cm}$$

$$L = 193 \text{ cm}$$

$$v = 1.7 \times 10^{-4} \text{ poises}$$

$$V = \frac{(3.14) \ (2666.7) \ (7.62)^4}{(1.28) \ (193) \ (.017)} = \frac{1764425.735}{26.248}$$

$$V = 67221.34 \text{ cm}^3/\text{sec}$$

(Table 6 provides metric conversion. Figure 7 provides a quick approximation of vapor-line conductance.)

TABLE 6.—METRIC CONVERSION

1 in	002.5400 cm	025.4000 mm
1 ft	000.3048 m	000.0304 cm
1 in^2	006.4520 cm^2	645.2000 mm^2
1 ft^2	929.0000 cm^2	
1 in^3	016.3870 cm^3	016.3870 ml^3
1 ft^3	000.0283 m^3	000.2830 cm
		.283 ml
10 mm	1 cm	0.3937 in
010 cm	1 dm	3.9370 in
100 mm^2	1 cm^2	.155 in^2
100 cm^2	1 dm^2	15.52 in^2
010 ml	1 cl	0.3380 fl oz
010 cl	1 dl	6.1025 cu in
010 dl	1 l	1.0567 fl qt
010 l	2.64 gal	

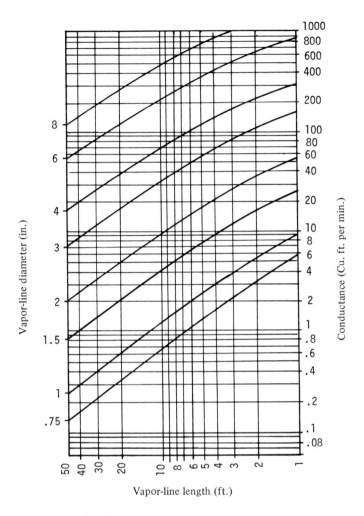

Figure 7. Vapor Line Conductance.

Vacuum Measurement

As it applies to freeze-drying, vacuum measurement falls within pressures ranging from 2 to 0.010 mm Hg (2000 to 10 μm Hg). Of the impressive array of available vacuum gauges, some are limited or impractical for freeze-dry application but there is a large variety from which to select.

Most vacuum measurement systems may be classified as mercury, mechanical, or viscosity manometers; radiometer or thermocouple vacuum gauges; ionization gauges; or resistance gauges.

Mercury Manometers. Mercury manometers range from highly sophisticated to simple barometers. The Zimmereli manometer is a complex U-tube with a pressure range from 0 to 300 mm Hg. While it deserves mention as a roughing gauge that may be used in tandem with another type, it does not operate in the ultimate range that is suitable for freeze-drying. The same may be said of the Dubrovin gauge with a range of 0.2 to 20 mm Hg.

The most satisfactory mercury manometer is the tilting McLeod gauge. It is precise, with readings from 5 mm to 0.005 mm Hg. The operating principle of the McLeod gauge is that with the compressing of a given volume of gas, whose pressure (P) is being measured, to a smaller volume (v) and observing the pressure (p) given by the relationship, $p = PV/v$ in accord with Boyles Law. As the gauge is tipped, mercury flows from the reservoir into a capillary tube; this serves as a reference tube, open at the top to the low pressure side of the system. Simultaneously, mercury flows into a closed tube of equal dimensions where the volume of gas is compressed and measured against a calibrated scale. The pressure measured is that at the instant that the gauge is tipped. The reading is not continuous; the gauge must be tilted and the reference column adjusted for each reading.

The McLeod gauge is easily cleaned—but it is glass, and breakable. It must be connected directly to the vacuum system and cannot be moved about as conveniently as an electronic-type gauge. The McLeod gauge does not indicate the partial pressures of condensable vapors, such as water, oil, or ammonia. Generally, however, neither oils nor ammonia vapors are involved at temperatures used for freeze-drying. At $-40°$ C. the vapor pressure of water is 0.096 mm Hg, which permits little or no water vapor to be presented to the gauge. Despite these considerations, the McLeod gauge does not have idiosyncrasies of many electronic gauges, does not require special calibration, and gives a positive reading within its operating range.

Mechanical Manometers. For many years efforts have been made to develop low-pressure measurement devices which depend upon mechanical deformation of a thin metallic diaphragm or spiral tube. These elements are hermetically sealed and mechanically linked to a movable pointer over a calibrated scale. No mechanical manometer has been found that is suitable for freeze-dry use.

Radiometer Vacuum Gauge. There are at least a dozen variations of the radiometer-type vacuum gauge, which was one of the earliest devices for detecting low gas pressures. (Sir William Crookes devised it in 1873.) The instrument is a glass bulb in which is suspended a four-plate vane mounted at a right angle to a center axle which is free to rotate. The vanes are black on one face and reflective on the other. When placed near a heat or light source, and when the pressure in the bulb is reduced, the vane rotates. As pressure is further reduced, the velocity increases. When extremely low pressures are attained, the rotation nearly stops. Knudsen applied the radiometer principle to a torsion-type gauge, and many modifications have been made on the Knudsen gauge. The principle has been applied from the primitive Crookes device to a highly sophisticated series of specialized laboratory measuring systems but none is recommended for freeze-dry application.

Viscosity Manometers. There is little to be said here about the viscosity manometer except that it exists and that it serves to measure small differences in pressure under controlled laboratory conditions but introduces significant error at low pressures. The operating principle is based upon the fact that the coefficient of slip of viscous materials is proportional to its mean free path and consequently inversely proportinal to the pressure. The viscosity manometer is not suitable for freeze-dry applications.

Ionization Gauges. The ionization gauge deserves mention primarily because it is not of use in the freeze-dry technique. It operates in pressure ranges from 10 μm to 2×10^{-7} mm Hg.

The operating principle is based upon the fact that when a charged electron passes through a gas, there is a definite probability that collisions between electrons and gas molecules will result in the formation of positive ions. The relatively high-velocity electron colliding with a gas molecule drives an electron from the molecule, which becomes positively charged. The number of positive ions produced at low pressure is directly proportional to the pressure of the gas. A measure of ion current then gives an indication of gas pressure.

Thermocouple Vacuum Gauges. The thermocouple gauge is probably the most efficient electrical gauge available. It is designed for reading total pressure of condensable vapors and permanent gases within the range of 3000 to 1 μm Hg. It consists of a thermocouple in proximity with a tungsten wire heater. Pressure is determined by the measured variation in the transfer of heat as a function relative to the variation in pressure within the system. This measurement is read directly in μm Hg on the scale of the gauge.

It may be purchased with precalibrated gauge control and matched gauge tubes, thereby making it possible to obtain pressure readings at any point in the system, without recalibrating the control system and by merely switching probe connections from place to place.

Resistance Gauges. The most typical is the Pirani gauge, which directly measures the pressure of air within the range of .25 to 250 μm Hg (.025 to .20 mm Hg). It is easily calibrated to the measurement of other gases (including small amounts of water vapor that occur in freeze-dry work).

The operation of the Pirani gauge is based on the effect of temperature on the change of electrical resistance in a wire that is sealed in an external glass element. Molecules within the system striking the resistance wire conduct and remove heat in accordance with their numbers and their specific heat. The resulting cooling of the wire brings about a change in resistance which is proportional to the pressure in the system. The resistance change, measured by an electrical circuit, is read directly on a meter calibrated in μm Hg.

The Pirani gauge is temperamental—easily affected by temperature changes and "ethereal currents" in the laboratory.

Refrigeration is primarily the application of thermodynamics to an understanding of the physical properties of refrigerants, and a knowledge of the mechanics of refrigeration systems. The first recorded mechanical refrigeration device was a hand-operated ether-vaporization system; the patent was issued in 1834 in Great Britain to an American, Jacob Perkins.

For freeze-drying applications, it is desirable to understand terms of measurement, economic feasibility of various systems, and the limitations imposed by the physical capacity of a compressor or expansion system. For example, temperatures near absolute zero may be achieved mechanically but the apparatus is costly, and the net gain in freeze-dry efficiency negligible.

Refrigeration deals with the absorption of heat from a given location, its subsequent transfer to another location, and the dissipation of that heat. Regardless of the conditions surrounding this transfer of heat, the laws of thermodynamics are involved. In the most commonly applied techniques, the refrigerant changes from liquid to vapor and is recondensed to a liquid. The thermodynamics of liquid-vapor phasing are complex.

The term vapor, as used here, is defined as a fluid that exists in a gaseous state near its saturation temperature at a given pressure. Ideal gas laws do not apply to vapors, a fact that becomes more evident as vapor saturation approaches the critical isotherm (the lowest temperature above which a substance can still be a liquid).

For freeze-drying, the most practical form of refrigeration is a compression system which incorporates recovery of the refrigerant. This type of equipment may be broadly classed as a compression system. Regardless of whether the compression cycle is produced by vaporization of a secondary fluid (absorption) or with rotary, centrifugal, or reciprocating apparatus, it is still fundamentally a compression system.

The most pertinent of these systems for our purposes involves the refrigerant alternating between the liquid and vapor phases, where the heat of vaporization is used to transfer heat from one place to another. It is important that a constant temperature be maintained.

The hermetically sealed refrigeration compressor is composed of a motor and compressor shaft of one-piece construction. The motor (cooled by the flow of the refrigerant gas) and compressor assembly are within a gas-tight housing that is welded shut. This construction eliminates certain parts (pulley, belt, compressor flywheel, and compressor seal) found in an open compressor unit—and the need to service and replace those parts.

The potential danger of a hermetically sealed unit is that under some freeze-dry conditions insufficient circulation of the refrigerant could cause the motor to burn out. This has not happened at the Smithsonian, where sealed units have been continuously used for eight years at ambient temperatures of 90° F. Our refrigerant was R-22; however, refrigerants with different operating characteristics might not fare so well.

Overheating could be corrected by installing a water-cooled condenser in the discharge side of the compressor.

A second type, the air-cooled compressor, usually operates with a belt and pulley; the motor is open, and cooled by air circulating around it.

Operating characteristics of both types of compressors are otherwise essentially the same.

Selection of a compressor should be based upon calculation of heat load on thermal insulation (see page 45).

Refrigerants

No compound is ideal under all working conditions. Two that are suitable for freeze-dry work are R-22 and R-502, both halogenated hydrocarbons. (See figure 8 for their operating characteristics.)

Condenser and evaporator operating pressures should be positive without going too high above atmospheric pressure. Negative pressures that may occur in the suction side of the evaporator permit leakage of air and moisture into the system and make it difficult to detect leaks. Extremely high operating pressures increase power consumption and require heavy construction of the compressor, evaporator, and condenser—and cause leaks. The only compromise that we have made in the selection of refrigerants is that R-502, under certain conditions, does operate below atmospheric pressure. In dealing with this problem, an extremely tight system should be built—with careful workmanship—and tested under high pressure. In addition, a water-cooled condenser should be used to control head pressure and produce a resultant increase in back pressure.

The toxicity of R-22 and R-502 is essentially the same. Both are considerably less toxic than dichloroethylene, and slightly more toxic than propane or butane; R-12 is even less toxic than R-22 or R-502. (Before spilling large amounts of these compounds into the room, refer to the Underwriters' Laboratories Classification of Comparative Life Hazard of Gases and Vapors, table 13.) Neither compound is flammable.

R-22, $CHClF_2$ boiling temperature $-40.8°$ C, is used in all types of refrigeration with reciprocating compressors. The thermodynamic properties of R-22 permit the use of smaller equipment than is possible with similar refrigerants.

R-502 is an azeotropic mixture, composed of 48.8 percent R-22 and 51.2 percent R-115, $CClF_2 - CF_3$, (by weight). It permits attainment of the capacity of R-22 with discharge temperatures of R-12 in reciprocating compressors. This refrigerant is finding new use in low temperature work.

R-12 dichlorodifluoromethane CCl_2F_2, which boils at $-81.4°$ C, is the most widely used Freon compound generally applied to reciprocat-

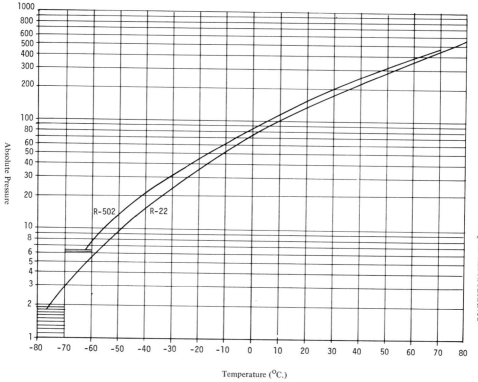

Figure 8. Refrigerant characteristics.

ing compressors. Its widest use is in air conditioning, but is may be used in freeze-dry refrigeration (see page 51).

(See temperature-pressure relationships of various refrigerants: table 14, page 175.)

Expansion Valves

For maximum vaporization of the refrigerant, it is important to select thermostatic expansion valves of correct capacity. And the valves must be installed at the proper locations. Such factors can spell the success or failure of the entire system. Thermostatic expansion valves for the condensing chamber should be a type designed to control temperatures below −40° C.

Although the valves in either chamber may be mounted in any position, they should be as near the evaporator inlet as possible. In the Smithsonian system they are mounted inside the chambers, where maximum efficiency is gained.

For satisfactory expansion-valve control, good thermal contact between the bulb and the suction line is essential. The bulb, which controls the expansion valve, should be fastened securely with two metal straps to a clean section of the suction line inside the chamber. The bulb should be near the midpoint of the line around the coil—not near the bottom of the line because a refrigerant-and-oil mixture is usually present there and would result in incorrect control of the expansion valve.

The filter and drier should be installed in the liquid line between the compressor and the thermostatic expansion valve.

Further protection is easily and inexpensively provided with a sight glass through which the refrigerant level can be checked. Bubbles in the liquid line indicate that the refrigerant level is low.

Thermal Insulation

Insulation material is required to minimize heat leak, which loads the refrigeration unit. This material may be glass wool (with a K factor of 0.29), rock wool (K factor = 0.26), compressed cork (K factor = 0.30), or similar material. The Smithsonian uses foam-in-place plastic (polyurethane). Foam-in-place plastics range in density from 2 to 10 lb per cu ft and offer K factors from 0.02 to 0.24. These plastics can be used without great difficulty and can be made to conform to any contour (table 7), but some are highly flammable.

Calculation for the determination of thermal impedence and heat load is as follows:

K = thermal conductivity factor (BTU/hr)

S = surface area of insulation (sq ft)

A = ambient temperature (difference between the inside and outside of chambers, in °F)

IT = thickness of insulation (in)

$$\frac{(K)\ (S)\ 2A}{IT} = BTU/hr$$

TABLE 7.—THERMAL CONDUCTIVITY OF VARIOUS INSULATING MATERIALS

Material	Density[1]	K[2]
Kapok (between paper)	1.0	0.24
	2.0	0.25
Balsam wool (chemically treated)	2.2	0.27
Linofelt (flax fibers)	4.9	0.28
Thermofelt (hair and asbestos)	7.8	0.28
(jute and asbestos)	10.0	0.37

TABLE 7.—THERMAL CONDUCTIVITY OF VARIOUS INSULATING MATERIALS (cont.)

Material	Density[1]	K[2]
Rock wool	6.0	0.26
	10.0	0.27
	14.0	0.28
	18.0	0.29
Glass wool	4.0	0.29
	10.0	0.29
Corkboard (rigid)	5.0	0.25
	7.0	0.27
	10.6	0.30
	14.0	0.34
Rock cork (rigid rock wool)	14.5	0.326
Insulite (rigid woodpulp)	16.2	0.34
Cellotex (rigid cane fiber)	13.2	0.34
Polyurethane (rigid sheet or foam in place)	2.0	0.02
	4.0	0.06
	6.0	0.11
	8.0	0.165
	10.0	0.24

[1]Density is given in pounds per cubic foot.
[2]K = Thermal conductivity in B.T.U. per hour per square foot at a temperature gradient of 1 F° per inch of thickness.

Temperature Measurement

Scientific suppliers offer a variety of temperature-measuring devices which range from simple thermometers to sophisticated remote reading devices. The two systems recommended are the telethermometer and the thermocouple gauge.

The telethermometer—a wide-span multirange transisterized thermistor system—is usually accurate within 0.5° C and is readable to 0.2° C. The thermistor probe, available in various shapes and sizes, is nothing more than a resistor made of a material whose resistance varies sharply in a known manner with temperature. The variation in temperature is related directly to the variation in current flow from the energy source in the meter housing. Current variation is read, in degrees centigrade, on a dial-type meter. Most telethermometers operate on batteries, which must be replaced after each 2000 hr of operation. Temperature ranges from −80° C to +40° C are available, with a disk probe attached to the surface of the specimen chamber or the condenser coil.

Thermocouple gauges are available for measuring virtually any temperature range. A thermocouple is a union of two conductors of dissimilar metals joined at their extremities. Electricity is produced by the direct action of heat and its unequal conduction. Simple readout meters attached to the thermocouple leads are read on a scale calibrated in degrees Centigrade. There are no batteries, and the meter is easily calibrated. See table 8.

TABLE 8.—CALIBRATION FOR THERMOCOUPLES
Iron versus Copper-Nickel (Iron-Constantan)

Degree C	Electromotive force given in absolute millivolts (Reference Junctions 0° C)									
	0	1	2	3	4	5	6	7	8	9
−40	−1.96	−2.01	−2.06	−2.10	−2.15	−2.20	−2.24	−2.29	−2.34	−2.38
−30	−1.48	−1.53	−1.58	−1.63	−1.67	−1.72	−1.77	−1.82	−1.87	−1.91
−20	−1.00	−1.04	−1.09	−1.14	−1.19	−1.24	−1.29	−1.34	−1.39	−1.43
−10	−0.50	−0.55	−0.60	−0.65	−0.70	−0.75	−0.80	−0.85	−0.90	−0.95
0	0.00	−0.05	−0.10	−0.15	−0.20	−0.25	−0.30	−0.35	−0.40	−0.45
+00	0.00	0.05	0.10	0.15	0.20	0.25	0.30	0.35	0.40	0.45

At 100° C the Emf is 5.27 mv

Degree C	Copper vs Copper-Nickel (Copper-Constantan)									
	0	1	2	3	4	5	6	7	8	9
−40	−1.48	−1.51	−1.54	−1.58	−1.61	−1.65	−1.68	−1.72	−1.75	−1.79
−30	−1.21	−1.16	−1.19	−1.23	−1.26	−1.30	−1.33	−1.37	−1.41	−1.44
−20	−0.76	−0.79	−0.83	−0.87	−0.90	−0.94	−0.98	−1.01	−1.05	−1.09
−10	−0.38	−0.42	−0.46	−0.50	−0.53	−0.57	−0.61	−0.65	−0.68	−0.72
0	0.00	−0.04	−0.08	−0.12	−0.15	−0.19	−0.23	−0.27	−0.31	−0.35
+00	0.00	0.04	0.08	0.12	0.16	0.20	0.23	0.27	0.31	0.35

At 100° C the Emf is 4.277 mv

*From National Bureau of Standards Circular No. 56.

The only aspect of the system that requires attention is the selection of thermocouple materials: copper-constantan is effective for temperatures from −130° C to +40° C; iron-constantan is recommended for −100° C to +500° C. The only disadvantage of the thermocouple gauge is that it may be temperamental when near other metal masses. Careful placing and shielding are required. For accurate temperature measurements, thermocouple lead wires must be of the same material as the measuring junction, or reference junctions must be installed and maintained at 0° C.

SIMPLE FREEZE-DRY APPARATUS IN A CHEST-TYPE FREEZER

The simplest apparatus for experimental freeze-drying is shown in figure 9. It is comprised of a deep-freeze chest, a specimen chamber or vessel, a vapor/vacuum line, and a vacuum pump.

The deep-freeze chest may be adapted as shown with a wooden

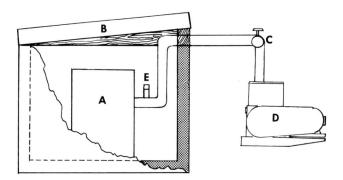

Figure 9. Simple Freeze-Dry Apparatus. *A*, specimen chamber; freezer cover (note wooden wedge); *C*, valve (optional); *D*, vacuum pump; *E*, air-admittance valve.

wedge spacer through which a hole is drilled, a hole large enough to accommodate the vapor line. Any attempt to cut through the side of the freezer to install the vapor line will probably lead to disaster, since most modern freezers are refrigerated with embossed aluminum sheet circuits. Standard deep-freeze chests are capable of maintaining temperatures of approximately −20° C, suitable for most freeze-drying.

The specimen chamber must be constructed to withstand external pressures up to 14.7 psi. These requirements may be met with a bell jar, a paint-spray pressure tank, or a pressure cooker. If a glass bell jar is selected, it must be shielded with hardware cloth to protect the operator in the event of implosion. If a pressure cooker is used, it should not incorporate a tapered metal-to-metal seal which may lock together permanently from the substantial external pressures that occur during evacuation of the system.

The vapor/vacuum line may be a large-bore rubber vacuum hose. See Vacuum Line Dimensions, p. 37. If the vacuum line is larger than the pump fitting, an adapter will be needed.

The vacuum pump must be gas-ballasted. Even with such protection, a certain amount of water vapor will accumulate in the pump oil, so the oil should be changed after every 40 hr of operation. The pump should not be allowed to stand for long periods when the oil is contaminated with water. After oil drained from the pump settles, it can be siphoned off and reused. After the vacuum pump has been drained, it should be flushed with clean oil.

The apparatus described is workable for freeze-drying. The next step in the evolution of a simple freeze-dry apparatus is the addition of a condenser, or cold trap.

Condenser

The condenser chamber must meet the same physical requirements as the specimen chamber. Within the condenser chamber, install a refrigerated coil (figure 10). Several configurations are usable: a coil within a coil, a similar arrangement made with finned refrigeration lines, or ready-made condensing cores such as those used in air-conditioning systems. The temperature of the condensing surface should be − 40° C to create a sufficient vapor-pressure difference to move water vapor from the specimen surface. The condenser also protects the vacuum pump from accumulating an excessive amount of water vapor. Oil should still be changed frequently—about once a month.

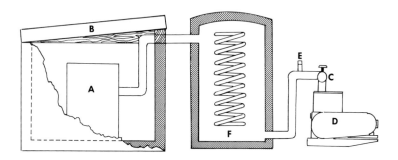

Figure 10. Simple freeze-dry apparatus with refrigerated condenser. *A*, specimen chamber; *B*, freezer cover; *C*, valve; *D*, vacuum pump; *E*, air-admittance valve.

The apparatus needs a valve to readmit air to the system. If a refrigerated condenser is used, the valve should be located so that air passes over the condensing surface before it enters the specimen chamber. If a condenser is not used, the air valve should be inside the freezer so that colder air re-enters the specimen chamber. In either system, the valve should be opened while the pump is operating; then the pump should be turned off, to avoid forcing it to run backwards or forcing the oil out of the pump and into the specimen chamber owing to higher external pressure.

FREEZE-DRY SYSTEM WITH A REFRIGERATED SPECIMEN CHAMBER

A 60-gallon paint-spray pressure tank is ideal for freeze-drying. Tank dimensions vary with the manufacturer but usually are 22 to 24 in in diameter and 30 to 36 in deep. Such a vessel—about the most economical on the commerical market—is already fitted with a cover, a cover gasket, and clamps which serve to achieve an excellent vacuum seal. The chamber may be mounted either horizontally or vertically. A horizontal chamber is easier to load and it may, conveniently, have open-mesh steel shelving. A vertically mounted chamber does not readily spill refrigerated air, thereby reducing the refrigeration load and low-temperature recovery time. The Smithsonian chambers are mounted horizontally for convenience in loading.

The vapor line is a piece of steel tubing, with a 4-in inside diameter, welded in an opening cut in the side of the chamber. The other end of the vapor line is fitted with a flange for later attachment to the condenser.

The chamber may be refrigerated by construction of a detachable refrigeration evaporator assembly (figure 11). The assembly is built over a light brass or copper sleeve which is fabricated 1 9/16 smaller in diameter than the inside measurement of the chamber. It must be fabricated so that an opening in the sleeve will be easily aligned with the vapor line where it enters the chamber. A continuous 5/8-in copper refrigeration line begins with a spiral coil filling the rear of the chamber and running the length of the outer surface of the sleeve. Straight tubing is joined at alternate ends with 2-in return bends to form the complete circuit. The tubing is soldered to the brass or copper sleeve.

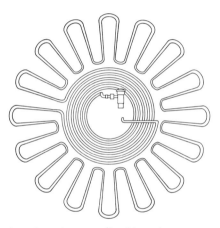

Figure 11. Configuration of continuous-coil refrigeration system. *A*, isometric view of detachable system; *B*, front view of detachable system.

Any attempt to solder the copper refrigeration lines to the zinc-coated steel chamber or to a steel sleeve is not recommended because the difference in the coefficient of expansion is great enough that a soldered union between such dissimilar metals will rapidly degrade and subsequently separate.

Once the assembly has been constructed, holes are drilled in the rear of the specimen chamber and the two refrigeration lines silver-soldered through the openings so that they enter just below the expansion valve (figure 12). Both lines, with flare nuts in place, should be flared with a flaring tool and the pressure line attached to the expansion valve (figure 13) as the assembly is installed in the chamber. The return line is then connected with a double-flare fitting to the second line in the rear of the chamber. At this point in the construction, the cover should be put in place with a good vacuum grease on the gasket and the chamber should be checked for vacuum leaks. To do this, a blank plate must be fitted on the flanged vapor line with a small tube sealed through the plate and attached to a vacuum pump. Pressure may be checked with a suitable gauge. The system may have to be pumped down several times and purged with air to clear the chamber of residual vapors characteristic of zinc and soldering flux.

This construction technique is suitable for almost any size specimen chamber.

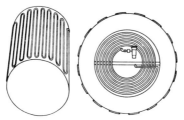

Figure 12. Detachable Refrigeration Coil.

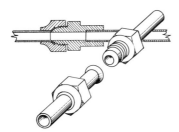

Figure 13. Flare Connector.

LARGE FREEZE-DRY SYSTEM

The largest refrigerated chamber of a freeze-dry system with which we have been involved is 5 ft in diameter and 9 ft long. Unless larger specimens are to be preserved, chambers this large are not recommended. It costs more for a system this large than for two or three smaller chambers. The larger the specimen chamber, the more complex the distribution of the refrigerant, and the less efficient the heat transfer.

In our oversize project we solved the problem of refrigerant distribution by dividing the refrigeration lines into six circuits operating from distributor heads attached to two expansion valves (figure 14). The chamber was refrigerated with a 2-hp water-cooled compressor. Refrigerant R-12 (see description on page 43) was used to chill the evaporator. (This chamber is so large that in the event of refrigeration failure sufficient freezer space is not available to preserve specimens being processed. As a fail-safe, a second compressor was installed and valved as a crossover standby (figure 15).) The vapor line is 6-in steel tubing fitted with a flange for connection with the condensing chamber. Finally, the outer surface of the specimen chamber was coated with a 2-in-thick layer of polyurethane foam plastic which was freon-blown.

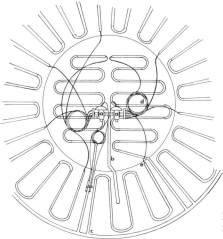

Figure 14. Refrigeration configuration in large specimen chamber.

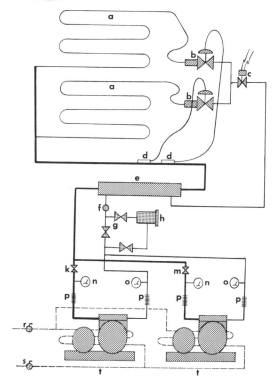

Figure 15. Schematic of large two-compressor freeze-dry system with crossover valves. *A*, evaporators; *B*, expansion valve; *C*, liquid-line solenoid; *D*, valve-control sensors; *E*, heat exchanger; *F*, sight glass; *G*, filter bypass and isolation valve; *H*, filter dryer; *K, M,* suction-line control valves; *N*, back-pressure gauge *O*, head-pressure gauge; *P*, vibration eliminators; *R*, cold-water inlet valve; *S*, water-waste valve; *T*, compressors. When *K* is open and *M* is closed, the left-hand compressor may be activated; to switch compressors, close *K*, open *M* and redirect power to right-hand compressor.

The condenser is of the type described on page 49 and shown in figures 16 and 17.

For a schematic of an entire freeze-dry system, see figure 2.

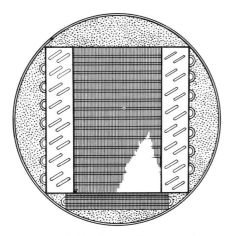

Figure 16. Refrigerated Condenser Core.

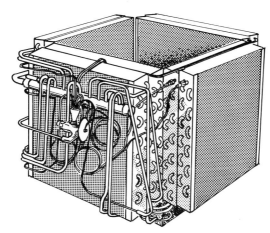

Figure 17. Refrigerated Condenser Core.

APPARATUS FOR FREEZE-DRYING CELL MATERIAL

The apparatus shown in figure 18 is designed specifically for freeze-drying cell materials and microscopic organisms for study under the scanning electron microscope. The specimen chamber—4 in in diameter and 24 in deep—is at the end of the vapor line (h) opposite the condenser (l). It is refrigerated with dry ice mixed with isopentane, which maintains a temperature slightly above −70° C.

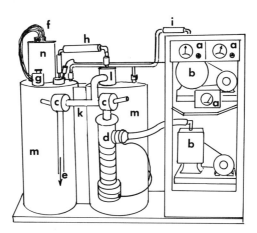

Figure 18. Cell-Drying Apparatus. *a*, vacuum gauges; *b*, vacuum pumps; *c*, quick-throw valves; *d*, diffusion pump; *e*, line to roughing pump; *f*, thermocouple bundle; *g*, probe selector; *h*, vapor line; *i*, Dewar evacuation line; *k*, vacuum line; *l*, condenser; *m*, Dewar canisters.

The system employs a cold-finger-type copper condenser refrigerated in a 1-liter bath of liquid nitrogen ($-195°$ C). Both chambers, in separate canisters, are insulated in vacuna (the Dewar effect) with a vacuum pump through line i. With a closed quick-throw vacuum valve over an operating diffusion pump, a roughing pump attached to line e, and the valve in line e open, the pressure in this system will be rapidly reduced to 25 μm Hg. As the pressure approaches 25 μm Hg, the quick-throw valves are reversed so that the diffusion pump is in the line. The pressure immediately drops below 1 μm Hg and continues to fall.

The specimen is frozen at liquid-nitrogen temperature and placed in the specimen chamber on a cover slip or an aluminum scanning electron microscope stub. The pump-down procedure is scheduled so that by the time the specimen temperature has approached $-70°$ C, the system will be at a pressure well below the vapor pressure of ice at $-70°$ C (1.97^{-3} mm Hg). Drying is rapid, and cytological integrity is maintained.

III

BIOLOGICAL APPLICATIONS

PREPARATION OF ZOOLOGICAL SPECIMENS

In museums, persons skilled in the taxidermist's art usually direct the freeze-dry preservation of exhibit specimens.

Most birds and mammals require washing[1] prior to processing. After the washing and drying, specimens should be rinsed with alcohol and insect repellent (see page 107).

To keep muscles from sagging as freezing progresses, the animal must be mechanically supported. Once specimens are in position, natural eyes are replaced with glass or plastic. Weight is measured and recorded periodically. In most cases the drying cycle is complete when the specimen ceases to lose weight. After rodents appear to be completely dry, however, the tail requires several more days of drying time; during this period the weight loss is so slight that it may be undetected.

Toward the end of the drying cycle, some specimens gain weight if new specimens are added to the chamber. These raise the vapor level in the chamber, and small amounts of vapor are absorbed by the slightly overdried specimens.

WIRE FOR MOUNTING SPECIMENS

Malleable, zinc-coated iron wire is recommended for mounting birds and mammals. To prepare it for use: secure one end of a 20-40 ft length in a vise (or fasten to a stationary object). Clamp the other end in the chuck of an electric drill, and twist under slight tension until the wire is straight and rigid. Then cut it into convenient lengths (approximately 2 ft) and with an electric grinding wheel taper one end of each to a point.

The size of the wire selected for each specimen should be sufficient to support the weight of the animal yet small enough to pass through

---

[1] Improper cleaning of feathers may ruin the specimen. See page 62.

the legs without distortion. Wire gauges and dimensions are listed in table 9.

A selection of a few sizes from 8 gauge to 22 gauge is satisfactory for mounting most birds and small mammals.

TABLE 9.—WIRE GAUGES AND SIZES
American or Brown & Sharpe

Gauge	Diameter (inches)	Gauge	Diameter (inches)	Gauge	Diameter (inches)
0	0.3247	14	0.0641	28	0.0126
1	0.2893	15	0.0571	29	0.0113
2	0.2576	16	0.0508	30	0.0100
3	0.2294	17	0.0453	31	0.0089
4	0.2043	18	0.0403	32	0.0080
5	0.1819	19	0.0359	33	0.0071
6	0.1620	20	0.0320	34	0.0063
7	0.1443	21	0.0285	35	0.0056
8	0.1285	22	0.0253	36	0.0050
9	0.1144	23	0.0226	37	0.0045
10	0.1019	24	0.0201	38	0.0040
11	0.0907	25	0.0179	39	0.0035
12	0.0808	26	0.0159	40	0.0031
13	0.0720	27	0.0142		

SMALL MAMMALS

Three wires are used to mount small mammals. The wires should be 4 or 5 in longer than the distance from the right front foot to the left rear foot when both legs are outstretched. Wires are pushed through the pads of the hind feet, into the hind legs, through the body cavity and out through the pads of the opposite front feet. The third wire is pushed from the anus through the body cavity under the spine into the base of the skull. The wires are bent in such a way that the animal stands erect. The leg wires are then pushed through holes drilled in the base, and the specimen is adjusted in a lifelike position. The wire protruding from the anus may be left long enough to support the head in the desired position. If abdominal muscles are sagging, the specimen may be suspended head down in the freezer chest and when it is semi-frozen, it may be shaped (claylike) into a lifelike position.

Since the entire animal is being used, other anatomical features are already accurately built in (figures 19 and 20).

An unposed specimen stored in a freezer too long may be partially dehydrated. When warmed to room temperature, rigid shrunken muscle masses may be restored by injection with a solution of 3-5 percent tribasic sodium phosphate. (At this time insecticides may also be injected.)

Figure 19. Flying Squirrel, *Glaucomys volans.* Dried at −20° C, 45 days.

Figure 20. Pigmy Marmoset, *Chichicos pygmaea.* Dried at −20° C, 32 days.

BIRDS

Techniques for positioning birds are similar to those for small mammals. Small-gauge, finely tapered wires are carefully pushed through the sole of the foot and along the leg bones without breaking the skin at the ankle joint. The wire is directed past the knee joint through the upper leg into the body cavity, then lodged beneath the shoulder blade or scapula. As with mammals, the third wire enters the body cavity through the anus, goes through the throat, and lodges in the base of the skull. The wires are then bent to position the bird. A soldering aid, such as shown in figure 21, is helpful in holding the head in position. Glass or plastic eyes are used, as in small mammals.

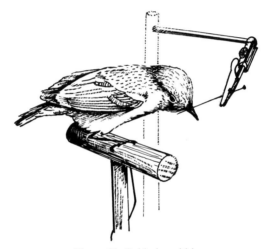

Figure 21. Soldering aid in use.

When final adjustments are made, care should be taken to avoid stretching the neck into an unnatural position. The folded wings are held against the body with long taxidermist pins. When posed in a flying position the wings may be supported with wires, or frozen at each joint with liquid nitrogen. Most birds and mammals will dry successfully at a temperature of $-20°$ C or below. (See figures 22–28.)

COLLECTING ZOOLOGICAL SPECIMENS

Many specimens brought to a museum's freeze-dry laboratory are road kills, or, in the case of birds, victims of picture windows or television towers. Other specimens, of course, are taken in the field with suitable equipment—shotguns, mist nets, traps. In all field collecting of biological specimens, compliance with Federal and State laws protecting wildlife is imperative.

Figure 22. Cedar Waxwing, *Bombycilla cedrorum.* Dried at − 15° C, 6 days (see drying chart, figure 138).

Figure 23. Yellow-breasted Chat, *Icteria virens virens.* Dried at − 20° C, 20 days.

Figure 24. Barred Owl, *Strix varia*. Dried at − 20° C, 130 days (see drying chart, figure 129).

Figure 25. Yellow-shafted Flicker, *Colaptes auratus*. Dried at − 20° C, 28 days.

Figure 26. Sharp-shinned Hawk, *Accipiter striatus velox*. Dried at −20° C, 36 days.

Figure 27. Robin, *Turdus migratorius*. Dried at −20° C, 24 days.

Figure 28. Starlings, *Sturnus vulgaris vulgaris.* Dried at −20° C, 20 days (average).

When animals are taken in the field, any large shot wounds should be plugged with cotton, as should the throat and anus. Each small bird should be placed beak first in a rolled paper cone with the feathers smoothed in natural position, and frozen as soon as possible.

Small mammals should be placed in plastic bags with fur smoothed in the direction of natural growth. Feet should be wrapped with cotton or paper towels so that the claws will not perforate the container.

Reptiles and amphibians may also be placed in plastic bags and frozen as soon as possible.

LABELS AND NOTES

As soon as possible after a specimen is obtained, a label should be made that includes the scientific name, locality, date, collector's name, size of specimen, eye color, and the colors of the feet, bill or facial skin.

Like all museum and study specimens, each should be labeled as completely and in as much detail as possible. Most of the world's laboratories have large collections of specimens that are virtually useless because they were improperly labled or not labeled at all.

CLEANING FEATHERS

The first opportunity to prevent bloodstains is in the field. Any blood on the plumage should be absorbed as soon as possible with sawdust or cornmeal, then daubed with cotton pads soaked with a saturated solution of bicarbonate of soda.

If the specimen is bloodstained when it reaches the laboratory, it can be washed with soap and water. If the stains persist, a saturated solution of sodium bicarbonate may be applied while the specimen is still wet—then rinsed with clear water after about 30 min. (A soft brush may help in removing a tenacious stain.)

Washing feathers requires special precautions to avoid damage, especially if the specimen was kept unfrozen for any length of time. Water should be tepid but not hot; mild, hard soap should be dissolved in water (see insect protection, page 107).

After the bird is washed and rinsed several times in clean water, it should be wrapped in paper towels to absorb as much water as possible. A final rinse in 90 percent ethyl or isopropyl alcohol will remove much of the remaining water and another wrapping in paper towels will remove most of the alcohol.

If the specimen is then rolled in hardwood sawdust or corn meal, any remaining alcohol will be absorbed, and the feathers will be restored. The absorbent material should be worked into the feathers and then shaken out.

The final step is to remove the minute particles of absorbent material from the feathers and dry the specimen—with a hose connected to a compressed air source, or in a rotating taxidermy tumbler with hardwood sawdust.

STOPPING BLOOD FLOW IN FRESH ZOOLOGICAL SPECIMENS

Occasionally during or after the washing process, blood will flow from the shot wound or the eye sockets. The flow may be halted by packing with cotton wet with strong formaldehyde solution.

In extreme instances where this is not effective, it is necessary to cauterize the offending blood vessel with a hot needle or the fine tip of a miniature soldering iron. Care must be taken to avoid searing the fur, feathers, or epidermis.

REPTILES

Most reptiles are easier to pose than birds or small mammals. Snakes rarely require supporting wires, unless the head is to be raised.

A snake may be draped over a tree limb or coiled on a base board, and the glass eyes installed; then it is weighed and frozen. If a snake is to be mounted with its head raised, as in a striking position, a soft iron wire should be inserted into the body through the mouth and bent to conform to the desired position. If the head rotates on the wire, a large pin can be inserted through the nostril and supported with a pin vice or soldering aid.

To mount a snake with its mouth open, the jaws are spread with a broken splint stick (as from a cotton swab).

Caution: Though the specimen is dead, the fang may retain enough venom to inflict a painful injury. There is also great danger of tetanus infection from snakebites; freezing the specimen does not necessarily reduce the risk of infection.

Fat leakage through the skin, causing discoloration, can be prevented by eviscerating the snake. Several small incisions across the belly between the ventral plates are adequate for removing the viscera and the accompanying fat tissue. If the specimen is fresh, it is not necessary to replace the viscera with cotton or tow.

Because some reptiles such as Gila monsters and monitor lizards have fat deposits in their tails that may eventually result in fat leakage, the fatty tissue should be removed after the drying process (see figures 29–32).

Although they take much longer to dry, turtles and lizards are treated much the same as small mammals. A turtle usually will dry more efficiently if ⅛-in or larger holes are drilled in the underside of the shell to accelerate escape of water vapor. Lizards will dry more quickly if their viscera are removed and replaced with taxidermist's tow or cotton, then closed with "blind" stitching.

Amphibians are posed in much the same manner as when mounted for casting. They are pinned to the mounting board to position the feet, and then a cut-cardboard support is placed under the chin, as shown in figure 33. They then are frozen, weighed, and freeze-dried.

Figure 29. Central American Rattlesnake, *Crotalus*. Dried at −20° C, 22 days.

Figure 30. Tabulated Tortoise. Dried at $-20°$ C, 132 days.

Figure 31. Alligator, *Alligator mississippiensis.* Dried at $-20°$ C, 9 months.

Figure 32. Monitor Lizard, *Varanus salvator*. Dried at −20° C, 56 days.

Figure 33. Mounted amphibian.

FISHES

Fishes, especially those that live in freshwater, are protected from loss of body salts by layers of mucus on the outer and inner surfaces of their skin. The inner (lipoprotein) layer impedes removal of water vapor from the specimen. Unlike the outer (mucoprotein) layer, which can be removed with soap and water, the inner layer is more resistant. Prior to freezing, after the fish is rinsed, it should be "fixed" in a 7 percent formalin solution.

The viscera and much of the tissue should be incised from the fish and replaced with taxidermist's tow or cotton. To mount a fish with its mouth open, while it's frozen, drill out as much tissue through the throat as possible without weakening the outer structure.

Despite these techniques, some fishes will not freeze-dry well. The only method that so far has proved successful (hardly an application of the freeze-dry technique) is to imbed the positioned fish in a block of plaster-of-paris which has been weakened by a small quantity of acetic acid (1 g per 100 ml of plaster). After the plaster has set, freeze the entire block. While the tissue is frozen, rout out as much as possible from the rear or nonexhibit side of the fish, as close to the skin surface as feasible. Then freeze-dry the entire block. When the fish is dry, coat the inner surface with catalyzed polyester resin, and sprinkle it with sawdust or short lengths of chopped fiberglass. Repeat this coating process until the specimen has a rigid internal support. (Resins are selected more by availability than by the specific requirements of this procedure. Polyester casting resins that are available from craft shops and resins in boat-repair kits are adequate for this process.)

All fishes require painting. The colors fade but the patterns remain, so the color may be restored with transparent oil pigments, such as those used in photo coloring (figures 34-35).

Fishes, amphibians, and reptiles are best dried at temperatures below −25° C, but amphibians should not be dried at a temperature higher than −30° C.

Figure 34. Pumpkinseed, *Lepomis gibbosis*. Dried at −25° C, 77 days.

Figure 35. Freshwater Fish. Dried at −25° C, 60 days.

MARINE INVERTEBRATES

For the most part, marine invertebrates are freeze-dried in much the same way as other specimens.

Crustacea, porifera, annelids, arthropods, and echinoderms may be freeze-dried with normal mounting techniques and at temperatures from −20° to −25° C.

Crustacea are mounted with the simplest mechanical supports possible, held in place with paper or cloth strips pinned to a mounting board. Drying, which is rapid, should be at a temperature of about −20° C.

All mollusks must be freeze-dried at −30° C; many of them, such as squid and octopuses, are best pre-fixed in 10 percent formalin, washed, and then freeze-dried. Mollusks that have been fixed for long periods and are misshapen from confinement in museum-jar storage are not suitable for freeze-drying. Specimens fixed for freeze-drying should be mounted first. Clams and oysters require no pre-fixing but should be freeze-dried at −30° C or below.

The killing process varies with the type of specimen and is seldom simple. Direct freezing, the wrong kind of anesthetic, or highly concentrated anesthetic will cause most clams, snails, slugs, oysters, and tubeworms to retract and thereby become useless as exhibit specimens. Anesthetics such as menthol, chlorobutonol, ethanol, and magnesium sulfate heptahydrate may be added drop by drop or crystal by crystal to the water surrounding the specimen until the specimen is relaxed; it can then be killed with a small amount of osmium tetroxide or formalin. The specimen must then be washed, frozen, and freeze-dried. Crustacea, porifera, annelids, or arthropods may be alcohol-killed, washed, and then freeze-dried. Most echinoderms may be alcohol-killed, but if a starfish is to be preserved with tube feet extended, it is necessary to quickly drop the specimen into liquid nitrogen for immediate freezing. We have successfully accomplished this with a starfish opening a clam. The finished specimen maintained perfect contact between the clam shell and the tube feet of the starfish.

Figures 36–41 show a variety of marine invertebrates freeze-dried at the Smithsonian Institution.

Figure 36. Marine Worms. Dried at −20° C, 10 days.

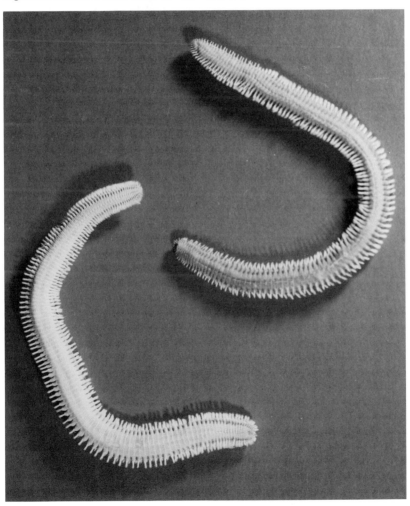

Figure 37. Hermit Crab, *Paguridae*. Dried at −25° C, 6 days.

Figure 38. Whale Louse, *Bathynomus giganteus*. Dried at −20° C, 8 days.

Figure 39. Blue Crab, *Callonectes sapidus.* Dried at −20° C, 6 days.

Figure 40. Ghost Crab. Dried at −20° C, 4 days.

Figure 41. Sponge. Dried at −25° C, 38 days.

EYES

"Eyes" are available from taxidermy suppliers for most game animals and many more common birds, but many kinds may be needed that are not commerically available. Small black eyes and pupils for larger eyes may be made from copper enameling glass heated with a torch, to form a small bead on a stainless steel plate (see figures 42, 43, and 44). The size of the glass bead is established by weight.

More complex eyes can be molded. Standard eye blanks are embedded in soft plastoline clay. The clay is coated with shellac, then liquid silicone rubber is poured over it (figure 45). Once the silicone mold is complete, eyes may be made directly with it. An alternative method uses a plaster or plastic cast from the silicone mold to form a plate. When the plate has polymerized, it will be used on a vacuum-forming machine to form a thin acetate mold into which the eyes are poured (figures 46 through 54). A thin layer of clear casting resin is poured in the bottom of each depression. Then a pupil (the round glass bead) is centered on the surface of the hardened plastic and another layer of clear casting resin is poured to support the pupil. After this layer has hardened, the color layer is poured to form the finished eye—or the clear surface is painted with colored resin to create the effect of a multicolored eye, and then a back-up layer of plastic resin is poured (see figure 48). Once the eye is complete, it is carefully cut out, leaving

Figure 42. Heat-Forming Glass Eyes, using Propane Torch.

Figure 43. Beading of Glass Chips.

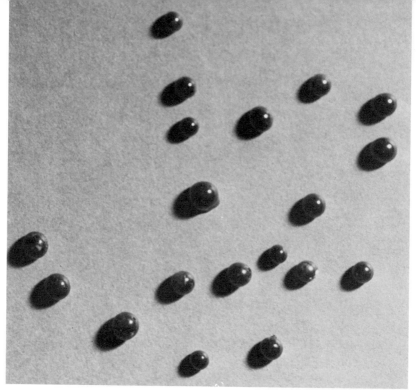

Figure 44. Heat-Formed Beads.

the acetate coating on the surface of the new eye. We have never had a problem of separation occurring between the acetate and plastic eye. The pupils can be formed in the same manner, eliminating the need to make glass beads.

Fish eyes may require different procedures because the pupil is not always round. In fact, there are many eye configuration, and each presents a special problem. It is, therefore, convenient to develop a number of molds, using the vacuum-forming technique. To do this, use an eye dropper to form plastic-resin master pupils on a sheet of glass, and select those with the most convincing appearance. The selected models are cemented to a small plastic block, and acetate may then be vacuum-formed directly to produce a large selection of pupils to be used as glass beads are used.

INSECTS AND INSECT LARVAE

For decades, tens of thousands of insect specimens have been perched near the heads of a variety of insect pins and dried in the normal room atmosphere without deleterious results. Many such insect collections are located in institutions of higher learning throughout the world.

In study collections, however, insect larvae and soft-bodied insects, as well as arachnids, are relegated to vials of alcohol, where they rapidly lose color and slowly deteriorate.

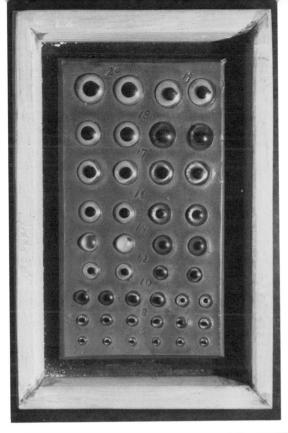

Figure 45. Eye Master.

Figure 46. Master Mold.

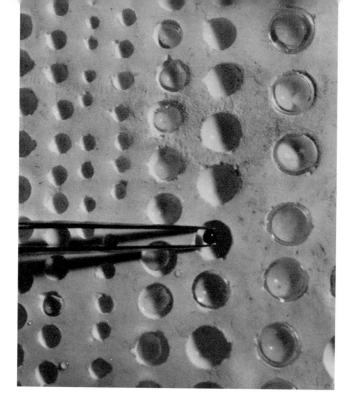

Figure 47. Placing Heat-Formed Glass Pupils.

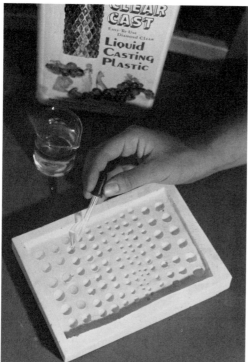

Figure 48. Pouring Plastic Eyes.

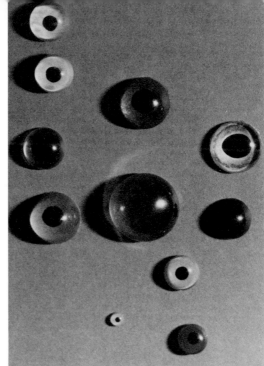

Figure 49. Finished Plastic Eyes.

Figure 50. Pouring Master
for Vacuum-Former.

Figure 51. Vacuum-Form Master.

Figure 52. Vacuum-Former.

Figure 53. Vacuum-Formed
Pouring Shell.

Figure 54. Eyes Poured in
Plastic Shell.

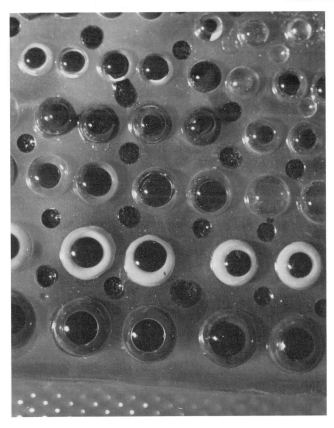

The advantages of the freeze-dry technique (or in some cases vacuum drying) for preserving these soft-bodied arthropods merit discussion even though the applications are primarily for exhibition, rather than for scientific study of the specimens.

Soft-bodied insects, insect larvae, and arachnids are best prepared with vacuum methods. M. S. Blum and J. P. Woodring are honing techniques of vacuum dehydration of insect larvae.

Most insect larvae and arachnids are small enough and have sufficiently porous cuticular structures that, when frozen, they permit an adequate flow of water vapor.

In fact, Blum and Woodring report excellent dehydration results at temperatures above freezing. My own experiments suggest that most of the water is removed while the specimen is frozen, and the temperature is maintained by the latent heat of sublimation. This is because the drying cycle is initiated while the material is frozen—and because the surface area of these animals is large in proportion to the mass.

It is also evident that if ice crystals melt in material with such a favorable osmotic gradient, there will be virtually no shrinkage because the water is so rapidly diffused that surface tension is insignificant (figures 55-58).

Figure 55. Black and Yellow Argiope, *Argiope aurantia.* Dried at $-20°$ C, 2 days.

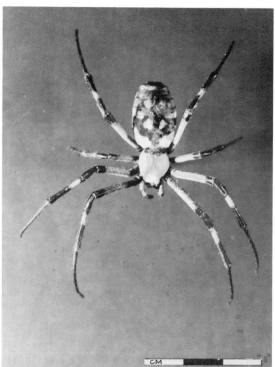

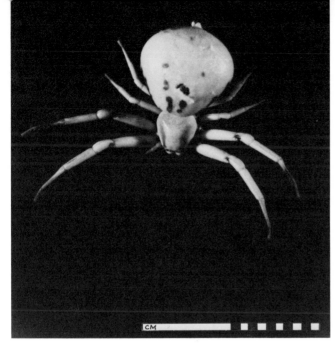

Figure 56. Archnid. Dried at − 20° C, 2 days.

Figure 57. Insect Larvae. Dried at − 20° C, 2 days.

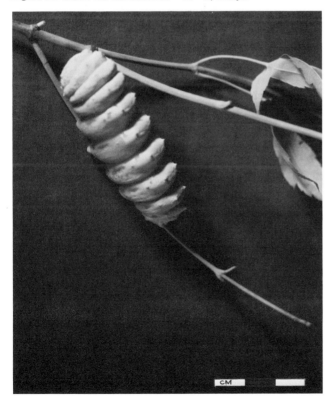

Figure 58. Praying Mantis. Dried at −20° C, 20 hours.

KILLING THE LARVAE

Since autolytic enzyme activity would be destructive to the specimen, Blum and Woodring kill the larvae by placing them in a freezer for 2 or 3 hr, positioning them when the specimen is semifrozen. Blum and Woodring advise against the use of insect pins during the freezing period: discoloration might result around the pins (possibly a result of oxidation of the polyphenols where the epicuticular barrier was perforated). If supercooling is a problem, the specimen must be killed before it is frozen in position.

KILLING ARACHNIDS

A solution of piperonyl butoxide is effective when sprayed as a fine mist over the specimen. This seems to have no adverse effect on color. As opposed to freeze-killing, the legs remain flexible, permitting the specimen to be posed in a lifelike position.

After freezing, arachnid specimens freeze-dry well at $-20°$ C in a vacuum chamber at a pressure of 60 μm or below.

Simple laboratory apparatus may be developed for both soft-bodied arthropod specimens and histology samples (figure 59).

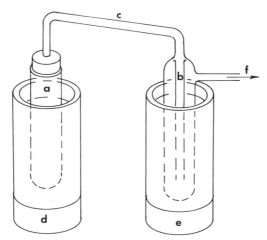

Figure 59. Simple histology apparatus. *a,* specimen tube; *b,* vacuum pumps; *c,* quick-throw valves; *d,* Dewar containers; *e,* line to roughing pump; *f,* vacuum tube to pump.

FRESH ANATOMICAL SPECIMENS

At this writing, Harris of the British Museum of Natural History has nearly completed the freeze-drying of a 6-kg adult elephant heart. At the Smithsonian Institution, in the National Museum of Natural History, the writer is freeze-drying the thoracic organs of an adult baboon.

In both laboratories, work with such materials has resulted in development of somewhat similar procedures, described here in the hope that readers will avoid our mistakes.

Although the baboon lungs (shown in figure 60) are described as fresh anatomical material, as distinguished from preserved material, they were collapsed and had been stored within the animal in a plastic bag in the freezer at $-20°$ C for about a year.

Procedure. The lungs, trachea, bronchi, heart, and pericardium were carefully removed as a single unit. The heart was separated from the pericardium, the pulmonary artery, veins, and aortic vessels were severed, the heart removed, and the pericardial membrane trimmed away.

A rubber tube was connected to the trachea with a small clamp and the lungs were expanded with air at a very low pressure. As they were expanded, the blood was expressed from the pulmonaries and washed away, and the whole apparatus placed in a freezer chest at $-20°$ C, with the lungs still under pressure from a small pump. The lobes were aligned as the tissue reached a semifrozen condition, and were left in the freezer for 24 hr.

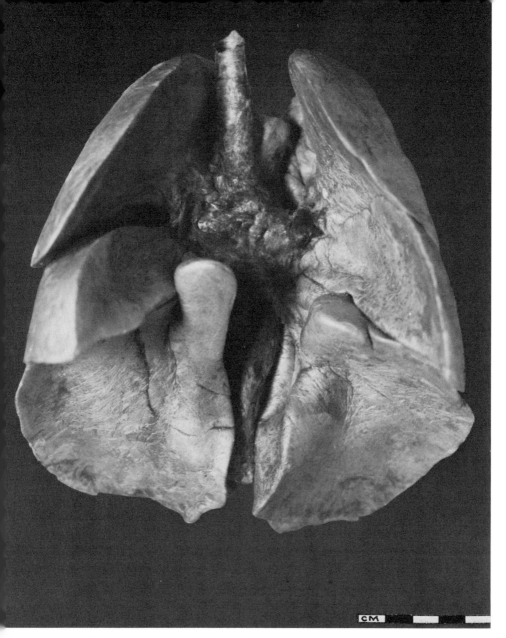

Figure 60. Baboon Lungs. Dried at $-30°$ C, 11 days (see drying chart, figure 113).

The lungs were then placed in the Edwards EF-2 freeze-dryer at $-30°$ C for eleven days. The total weight loss was 175.3 g or 66.7 percent. (For the histological preparation of lung tissue, see page 100.)

Figure 61 presents a typical specimen for anatomical study. With a jeweler's fine-bladed saw, the freeze-dried baboon heart was cut into five sections—providing a long-lasting study series which permits a better idea of structure. A series of animal hearts prepared in this way could serve the needs of a class in comparative anatomy for several years. If a sturdier specimen is required, such sections can be embedded in clear casting resin.

(Two other specimens worthy of note—a dissection of a human stomach, and a sliced section of normal human kidney—are shown in figures 62 and 63. Both specimens were frozen at −70° C; their temperatures were raised to −25° C, then freeze-dried in the Smithsonian chamber. The color has subsequently faded from both specimens, probably due to oxidation of the hematin compounds normally associated with colors dependent upon the presence of capillary blood.)

Figure 61. Sectioned Baboon Heart. Dried at −25° C, 21 days.

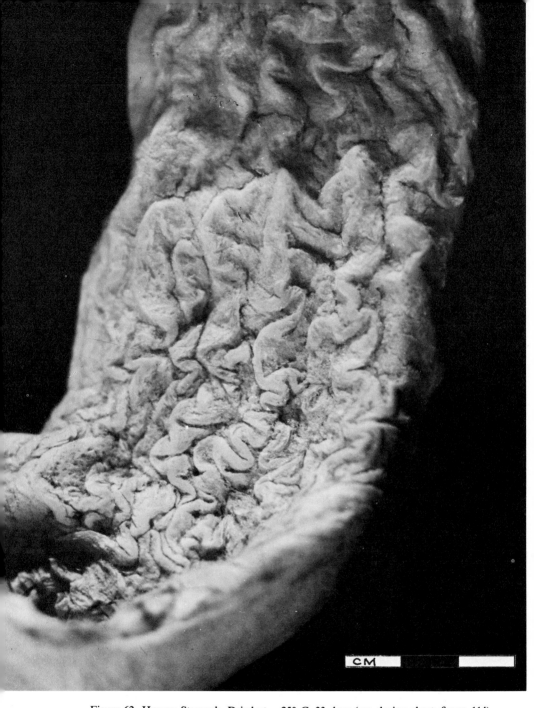

Figure 62. Human Stomach. Dried at −25° C, 33 days (see drying chart, figure 114).

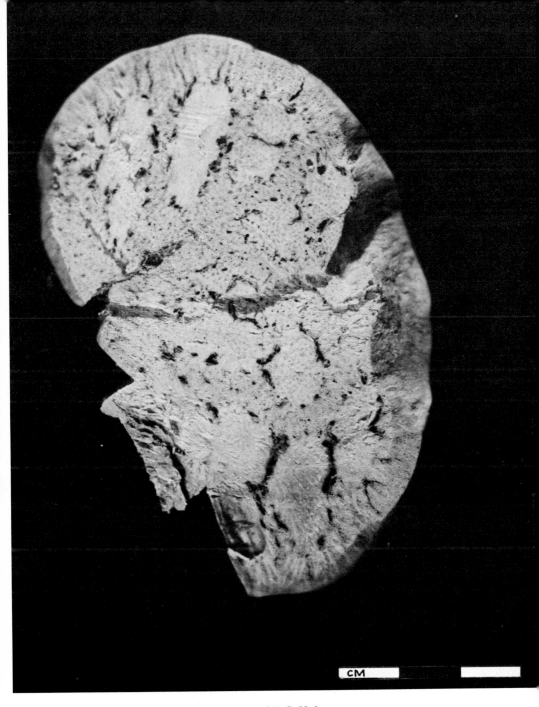

Figure 63. Human Kidney Section. Dried at −25° C, 28 days.

PRESERVED BIOLOGICAL SPECIMENS

Although the effects of freeze-drying previously preserved biological materials are highly variable, the only sources of many biological specimens are the already existing formalin-preserved collections. Certain formalin-fixed pathological specimens freeze-dry more satisfactorily than similar materials not treated with formalin.

The widespread applications of aqueous formaldehyde as a preservative for animal tissues are well known.

The chief difficulty in using formalin for preserving biological tissues arises from its chemical action. Typical of aldehydes, formaldehyde absorbs oxygen rapidly and is then oxidized to yield formic acid (pH ranging from 3.5 to 3). Such an acid solution in contact with biological materials produces a complex series of chemical changes that alter the formaldehyde solution, resulting in degradation.

To some extent, products of the denaturized protein and the decalcification of osseous materials neutralize formic acid, perhaps as high as pH 4, depending upon how long such materials have been in solution.

The decalcification of ossified material and esterification of fatty acids in formalin-preserved specimens further complicate the freeze-drying of such preserved materials. In addition, there is extensive precipitation of calcium phosphate, and the precipitate adheres tenaciously to the surface of freeze-dried specimens. Accordingly, to overcome disadvantages of formaldehyde preservation for study specimens as well as for freeze-dried exhibit specimens, it becomes necessary to provide a neutral formaldehyde solution.

Since a neutral or slightly alkaline formaldehyde solution is called for, the problem is how best to remove the acid. Too much alkalinity will alter the formaldehyde by self-condensation, making it of less value as a preservative. Therefore, we must select a method of neutralization in which the solution maintains a pH of 7–8.

Carbonates, bicarbonates, etc., have proved unsatisfactory for several reasons, but chiefly because they involve the continuous production of carbon dioxide.

Borax (5–10 g per l) is widely used as a neutralizing agent, but it leaves a deposit on freeze-dried specimens that makes it a liability. And it, too, promotes the self-condensation of formaldehyde.

Actually there are few suitable neutralizers available, for we are limited to those that are soluble in water and do not alter the preservative value of the solution.

A chemical that meets these requirements is hexamethylenetetramine (or hexamine), produced by a chemical reaction between formaldehyde and dilute ammonium hydroxide. The substance, crystalline in form, is water-soluble, and yields a weak basic or neutral solution (pH 7.6 to 8). While hexamine may be added to specimens that have been kept in formaldehyde for many years, the direct use of dilute ammonium hydroxide is equally effective.

To avoid a violent reaction when mixing a new solution, ammonium hydroxide should not be added directly to strong formaldehyde. The formaldehyde should first be diluted and the ammonium hydroxide added slowly, and stirred constantly. Specimens that have been stored in acid formaldehyde for some time should be treated in the neutralized bath for at least 24 hr.

Table 10 gives the quantities of ammonium hydroxide and formaldehyde required to produce neutral formaldehyde solutions at a variety of concentrations.

TABLE 10.—FORMULATION OF NEUTRAL FORMALIN

Percent formalin desired	40 Percent formalin (vol/liters)	Water (liters)	Ammonium hydroxide (milliliters)
3	1	6½	200
4	1	6	150
6	1	5	125
8	1	3	100

After obtaining a nearly neutral preservative solution, residual precipitates must be removed and low vapor pressure and eutectic solutions must be replaced with water.

Washing neutrally preserved specimens for 24 hr in slowly running tap water ordinarily makes them suitable for freeze-drying. This has been successful at the Smithsonian for pathology specimens such as sections of human brain (figure 64), human liver (figure 65), bovine eye tissue (figure 66), human spleen (figure 67, the upper specimen; the lower specimen was dried from a freshly frozen state), pre-fixed polycystic kidney (figure 68), lung material (emphysema—figures 69 and 70), and human fetus with intact uterus parts (figure 71). The fetus had been preserved for several years in an unbuffered formaldehyde solution, highly acid (pH 3), and then frozen in solution for about a year. The surface of the specimen was covered with a flocculent protein residue and calcium phosphate particles. The epidermal layer over the skull was slightly cracked and the soft tissue in the thorax, arms, and legs showed evidence of shrinkage.

Processing a Human Fetus

First, the fetus was washed in a large beaker under a gentle flow of cold water for 18 hr, then soaked for 8 hr in a solution of 30 ml of ammonium hydroxide added to 1 l of clear water. The specimen was then washed again with tap water for 24 hr.

The fetus was frozen at $-70°$ C overnight then placed in the Edwards EF-2 freeze-dry chamber regulated to hold the specimen at $-30°$ C, and the pressure in the system from 40-30 μm. There was virtually no change in the appearance of the specimen. The preserved fetus at the end of step 2 weighed 2,232 g. Its final dried weight is 162 g. In 7 wk the weight loss was 92.74 percent.

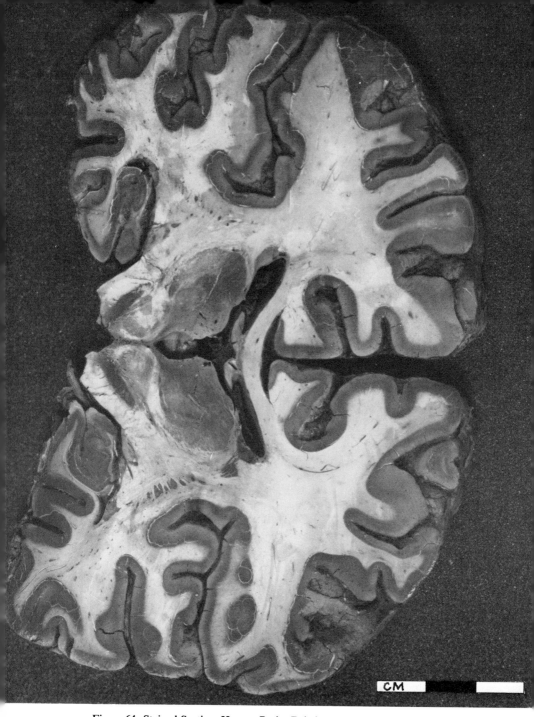

Figure 64. Stained Section, Human Brain. Dried at $-30°$ C, 15 days (see drying chart, figure 120)

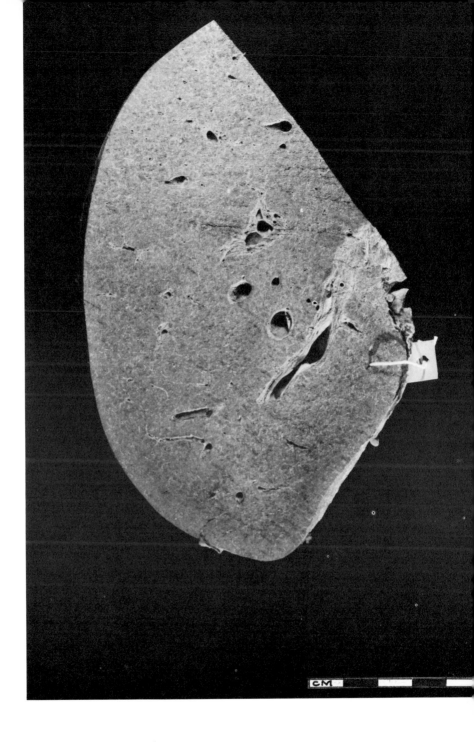

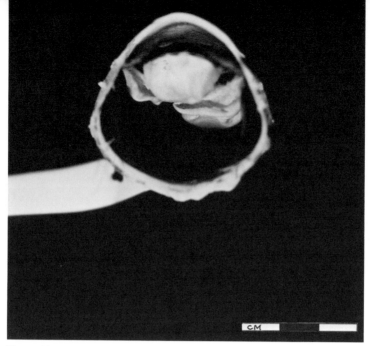

Figure 66. Bovine Eye Tissue. Dried at $-30°$ C, 20 days.

Figure 67. Human Spleen Tissue. Upper sample was first fixed with 5 percent formalin, dried at $-15°$ C, 14 days; lower sample was freeze-dried from a fresh specimen, dried at $-30°$ C, 12 days.

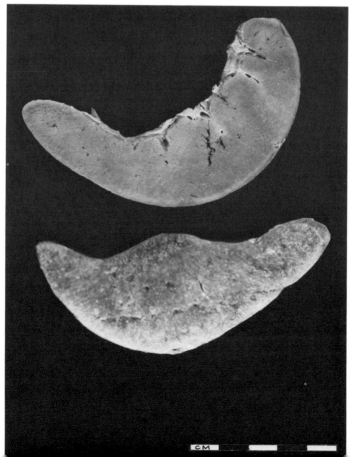

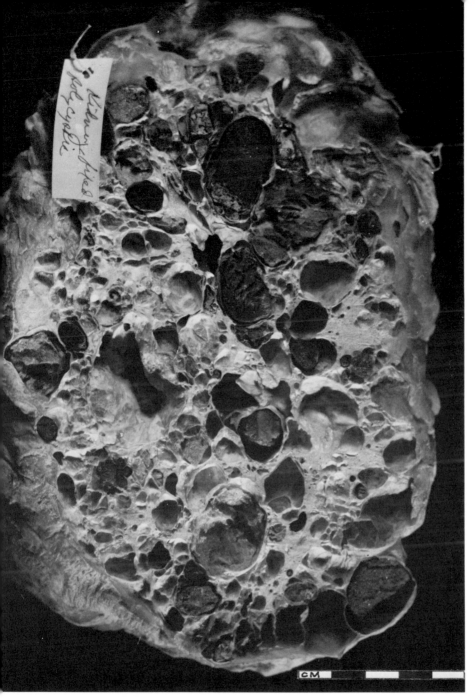

Figure 68. Human Kidney, Polycystic. Dried at −30° C, 24 days.

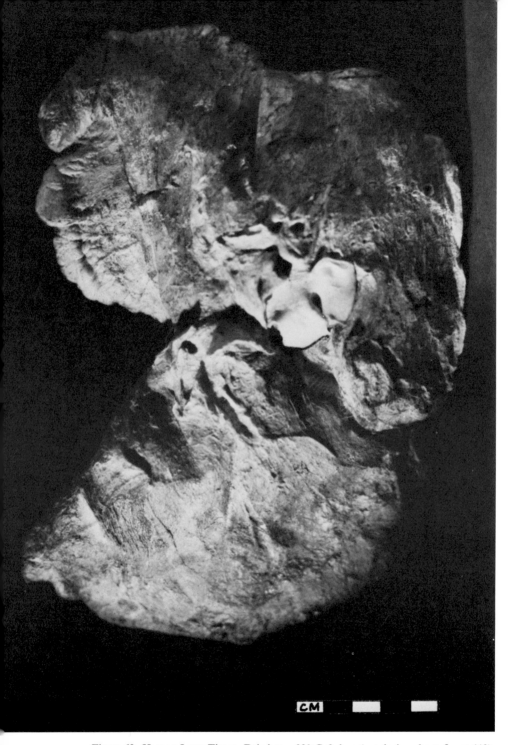

Figure 69. Human Lung Tissue. Dried at −30° C, 9 days (see drying chart, figure 119).

Figure 70. Human Lung Tissue. Enlarged section has surface removed to expose internal structures.

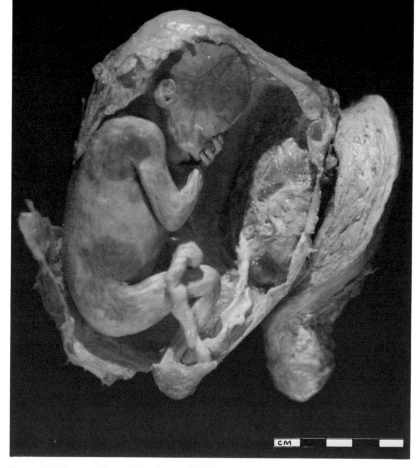

Figure 71. Human Fetus. Dried at −30° C, 7 weeks (see drying chart, figure 115).

Although the fetus might have dried much more quickly at a higher temperature, at −30° C the shrinkage of this already poor specimen was minimized, resulting in a usable study item.

Processing a Human Brain

Before the brain arrived at the Smithsonian laboratories, it had been pretreated by formalin fixation, slicing, and staining of some sections with Prussian blue to differentiate the cortex from the white material (figure 72).

In the Smithsonian freeze-dry laboratory the sections were placed in distilled water with a few drops of ammonium hydroxide to neutralize the formic acid (by a crude titration technique). The sections were then washed in running tap water overnight. A small section of tissue was checked with a universal pH indicator to ascertain that it is neutral.

After the brain sections were frozen at −70° C, they were placed in the Edwards EF-2 at −30° C for 14 days. The weight loss was 463.2 g, or 79.9 percent.

Figure 72. Human Brain Section, Medulary. Dried at −30° C, 14 days.

A few minute cracks developed in some of the massive sections, but no more than in specimens kept in formalin. By contrast, note cracks in surface shown by figures 73, 74, and 75. These specimens were dried at $-5°$ C.

Color contrast between the cortex and the white material increased as the water was removed.

Figure 73. Section of Human Liver. This specimen was formalin-fixed, washed out, then freeze-dried. Note cracks resulting from drying at temperature above $-5°$ C, (4 days).

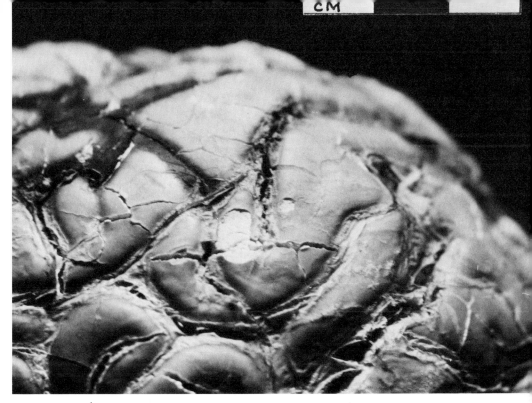

Figure 74. Human Brain Tissue. Note surface cracks resulting from a drying tempeta-
ture of −5° C (7 days).

Figure 75. Human Brain Section. Note surface cracks resulting from high drying temperature of −5° C, (6 days) in this stained section of formalin-fixed brain tissue.

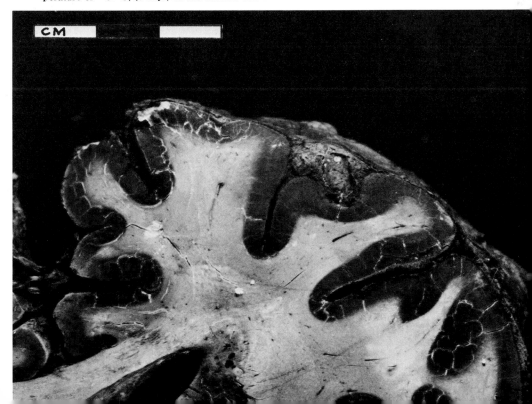

PATHOLOGY SPECIMENS

While the technique of freeze-drying pathologic material varies only slightly from other freeze-drying, some of our methods may deserve special mention.

Much laboratory work in pathology is with fresh or poorly preserved specimens. We recommend dissecting pathologic organs and tissues from their matrix and freeze-drying them for future reference. Freeze-drying is especially useful for specific types of pathologic materials. Human bone that contains soft cancerous growths cannot be sectioned without destroying the structure of the tumor. If the specimen is first freeze-dried, the sectioning procedure becomes simple and produces little damage. The humeral bone shown in figure 76 is a good example. Also unique by virtue of the technique is the heart shown in figure 77. The patient was a victim of arteriosclerosis with calcification in the heart valves. The xerograph (figure 78) shows the calcification as white blotches. If the heart is sectioned through a valve, a more exacting study can be made. Final pathological analysis of this specimen indicates a calcified stenotic mitral valve and a hypertrophied left ventricular wall (figures 79 and 80). This specimen will serve in a permanent collection for teaching and as a rare symptomatic reference.

Figure 81 is a rare example of lymphosarcoma involving the right ventrical, with a mechanical obstruction of the tricuspid valve. It was sectioned before freeze-drying. Many types of brain tumors may be freeze-dried and stored without change in structure. Figure 82 illustrates a matching tumor located within the skull and the depression in the brain. The skull is reversed so that the tumor and the depression are both left of the medial line.

MICROANALYTIC STUDY OF INFLATED WHOLE LUNGS

Most unfixed inflated lungs that are freeze-dried produce superior histological and cytological specimens. Further, soluble components such as mucoproteins, as well as metals and particulates, are still locatable. Freeze-drying lung tissue thus avoids the disadvantages of formalin-fume fixation and air drying.

The initial freezing of lung materials for microanalytic study is difficult; certain techniques are important. Due to the insulative nature of lung material, heat transfer must be rapid. Inflating the lungs with chilled helium is the best way to insure this. Helium gas is circulated at low pressure, through a simple coil suspended in liquid nitrogen, into the trachea and through the lungs. The lungs should be suspended in a freezer during this process in order to prevent thawing. Helium has a thermal conductivity factor of 6.09 times that of air, and 6.21 times that of nitrogen (see table 11).

No discussion of whole-lung preparation could be complete without mention of the work being done by Dr. Jerrold L. Abraham of the

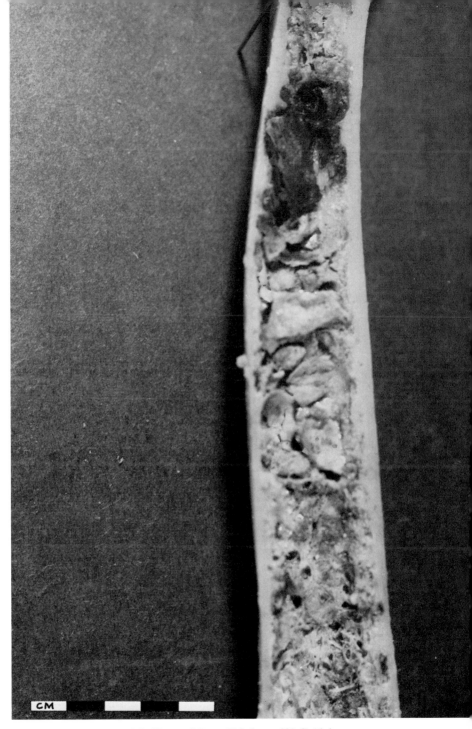

Figure 76. Cancer Growth in Humeral Bone. Dried at − 30° C, 16 days.

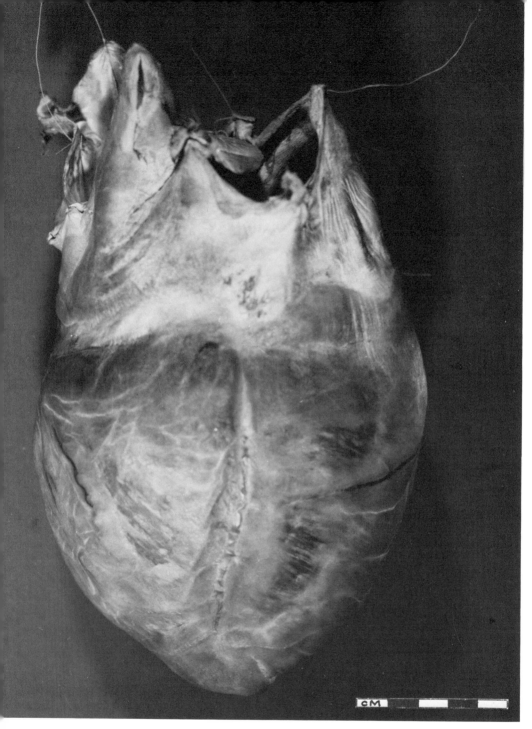

Figure 77. Enlarged Heart (Arteriosclerosis). Dried at −30° C, 23 days (see drying chart, figure 117).

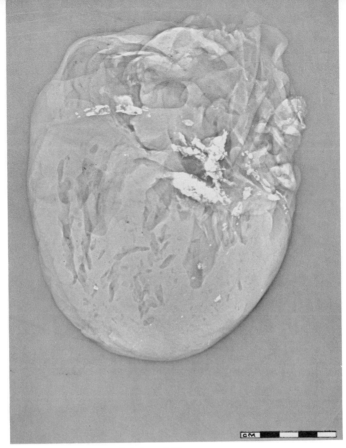

Figure 78. Xerograph of Enlarged Heart.

Figure 79. Section of Heart Valve.

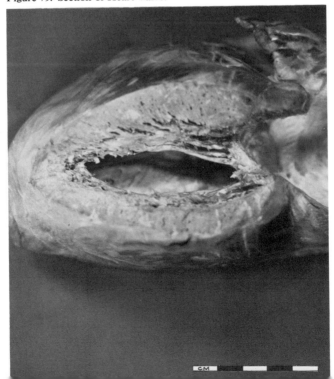

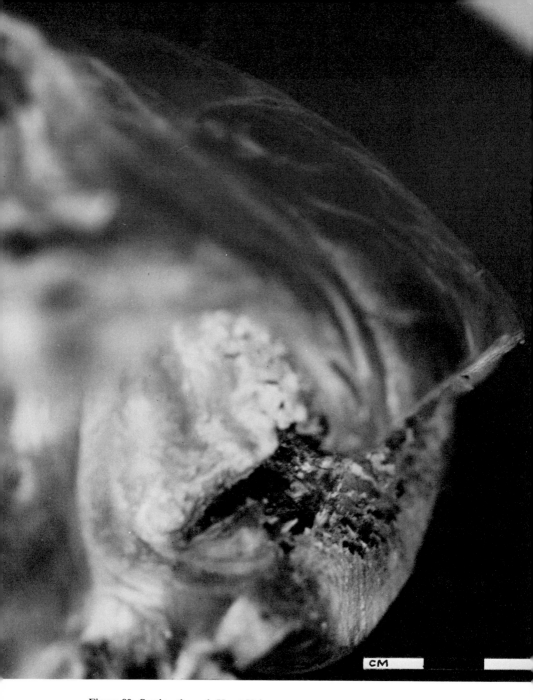

Figure 80. Section through Heart Valve.

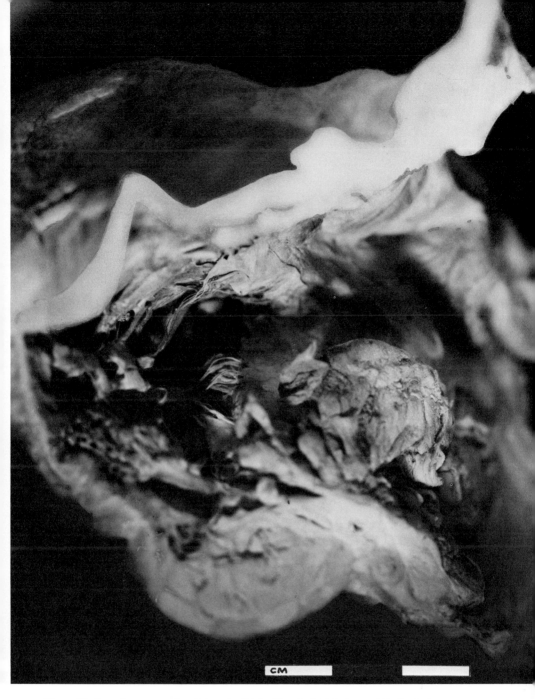

Figure 81. Lymphosarcoma Involving the Right Ventrical. Dried at −30° C, 12 days (see drying chart, figure 118).

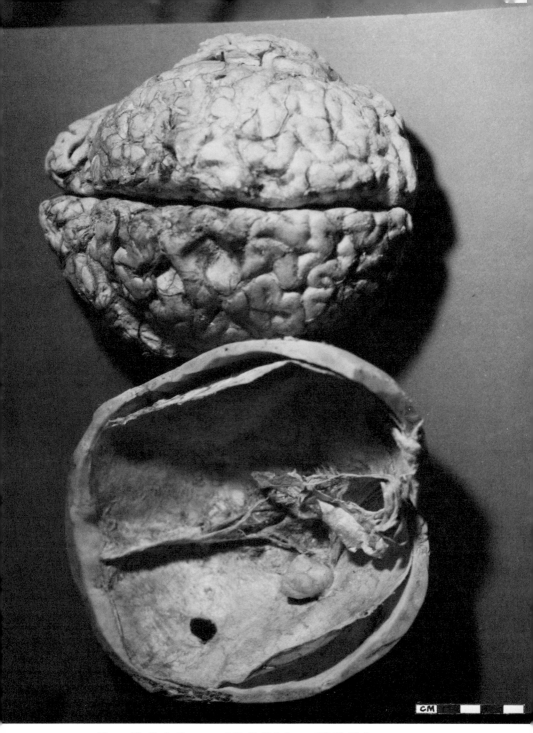

Figure 82. Brain Tumor and Skull. Dried at $-30°$ C, 16 days.

TABLE 11.—THERMAL CONDUCTIVITY OF GASES

T°C	Air	Helium	Nitrogen	Oxygen
−73.3	−	274.8	44.22	43.72
−40°	50.09	304.99	50.42	50.54
−28.9	52.15	314.49	52.48	52.81
−17.8	54.22	324.00	54.55	54.96
− 6.7	56.24	333.5	56.20	57.24
4.4	58.31	343.42	58.27	59.43

The values given are cal/(sec)(cm^2)(°C/cm) × 10^{-6}

Department of Pathology at the University of California (San Diego) School of Medicine and Dr. Phillip B. DeNee of the Appalachian Laboratory for Occupational Safety and Health (NIOSH) in Morgantown, West Virginia.

They are freeze-drying unfixed whole lungs for correlative roentgenological (X-ray), pathological, and microanalytical study of lung disease. Their studies involve roentgenography (X-ray photography) before and after freeze-drying. Fully inflated freeze-dried lungs are cut with a band saw into 2.5 cm slices and roentgenographed. These sections are selectively chosen to be serially sliced into thin sections on a rotating blade slicer.

The thin sections are photographed and roentgenographed. Areas are selected and cut from the thin slices randomly and through sections of gross lesions and analyzed for elemental composition by X-ray fluorescence, then processed for light microscopy, scanning and transmission electron microscopy, and electron-probe microanalysis.

Thus, the pathology and the amount and type of inhaled dust can be correlated with specific radiologic changes.

For this branch of pathology, the freeze-dry method is superior to any other preservation system. Freeze-drying retains soluble components as well as metals and particulates. It also provides greater crytological and histological integrity than the more standard formalin-fume fixation and air-drying methods.

PROTECTING SPECIMENS FROM INSECTS

Freeze-dried museum specimens are subject to infestation by a variety of insects, although—with proper treatment—probably no more so than specimens preserved by conventional taxidermy. The skins of conventionally preserved animals have more built-in protection because most are dusted with borax before they are mounted or stored.

Dermestid beetles—small oval insects usually covered with minute scales—lay their eggs in furs, hides, feathers, and, if the opportunity presents itself, museum specimens. Such materials are nourishment for the larvae. Infestations of the museum cabinet beetle (*Anthrenus musaeorum*), about 4 mm long in the larval stage and 2 to 3 mm long in the adult beetle stage, can destroy collections of insects, birds, and mammals in a very short time. Other insects causing damage are the carpet beetle (*Anthrenus scrophulariae*), the bacon beetle (*Dermestes lardarius*), and the common hide beetle (*Attagenus pellio*). Of course there are many others, but if the specimens are protected from these, they will be protected from others as well.

Many insecticides are effective against these museum pests, but arsenic or mercury compounds should be avoided because they are toxic to humans. Borax dusted into fur or feathers is of questionable value; primarily a dessicant, it seems to discourage only moths. Paradichlorobenzene and similar moth crystals are detrimental to both conventionally prepared and freeze-dried materials: the crystals are fat solvents and, even in vapor form, dissolve fats which will then flow to the specimen's surface: oxidation then damages the specimen.

The best insect-proofing agent that we have found is an aromatic sulfonamide derivative known as Edolan U.[1] It can be mixed with water or alcohol in any proportions, and resists oxidation and reduction. For protection against *Anthrenus* or *Attagenus* (and most other species of beetle pests) the specimen should be washed, then rinsed in a 1.5 percent[2] solution of Edolan U followed by a second rinse in a 1 percent solution of acetic acid. The specimen must then be washed in methyl alcohol to remove suplus Edolan. A specimen dried after the acetic acid bath may be washed in white gasoline (but beware, it's explosive!). An alternate method, especially if the specimen has not been washed, is to inject a 1.5 percent Edolan solution into the specimen prior to freeze-drying and to distribute the injection throughout the specimen. Or the specimen may be injected after freeze-drying with a 1 percent alcohol solution of Edolan and then air-dried.

Fur or feathers may be dusted with TMTD, Bis-dimethyl thiocarbamyl disulfide, sometimes called Tetramethyl thiuram disulfide. This agent is harmful if swallowed or inhaled, but less dangerous than other poisons or repellents. TMTD is also an excellent rodent repellent and fungicide.

SHIPPING FROZEN BIOLOGICAL SPECIMENS

Frozen specimens should be wrapped in newspaper, then packed with dry ice in a plastic-foam picnic cooler. The cooler in turn should be packed in a corrugated container with at least 2 in of crumpled newspaper all around. Ten lb of dry ice—packed close to the specimen at the last possible moment—will keep most specimens frozen for 48 hr long enough for air-freight shipment between any two points east or west of the Mississippi. Thirty lb of dry ice will hold the material for 72 hr, long enough for cross-country air-freight shipment.

---

[1] Available from Verona Corp., P.O. Box 385, Union, New Jersey 07083.

[2] Based on anticipated dry weight of the specimen.

Birds should be in paper cones, with their feathers pressed naturally against their heads and bodies.

All shipments should be labeled "Rush!—Perishable Frozen Biological Material."

PREPARATION OF SPECIMENS FOR SCANNING/ELECTRON MICROSCOPY

Unlike a transmission electron microscope, a scanning electron microscope scans the specimen surface and reads only the secondary emissions from the specimen. Only the surface of the object is examined. Minerals and metals require no pretreatment so long as they are clean and are electrical conductors.

Biological materials, however, must be specially prepared for viewing under the scanning electron microscope. They must be fixed, cleaned, dehydrated, and then coated with an electrical conductive material.

One technique is referred to as "critical-point drying," a method described by Thomas F. Anderson in the *Transactions of the New York Academy of Science* (1951). The method, stated briefly, is to fix the specimen, usually with osmium tetroxide, and pass it through a graded ethanol series to a clean amyl acetate and quickly place it in the bomb of the critical-point drying apparatus (figure 83). The bomb is then

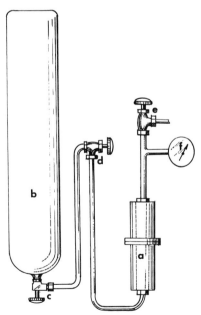

Figure 83. Critical-point drying apparatus. *a*, bomb; *b*, CO₂ cylinder; *c*, cylinder valve; *d*, control valve; *e*, release valve; *f*, pressure gauge.

flushed out with liquid carbon dioxide to replace the amyl acetate and refilled with liquid carbon dioxide. The temperature is raised to about 50° C., at which point the carbon dioxide passes its critical point and changes to a gaseous state. The valve is then opened slowly to allow the carbon dioxide to escape. The dried specimen may then be transferred to a metal specimen stub or glass cover slip which will be mounted on the stub. Standard electron microscope stubs are prepared to receive the specimens or cover slips with a thin layer of adhesive dissolved from transparent pressure-sensitive tape or "permount" mounting medium.

While early efforts in freeze-drying specimens for scanning electron microscopy were not always satisfactory, the efforts of many recent workers have been excellent, owing in large part to cleaner specimens, more care in technique, and freeze-drying at a lower temperature.

OSTRACODS

Ostracods, a subclass of Entomostraca—small crustaceans with unsegmented bodies—are enclosed in bivalve carapaces (shells), with locomotion provided by jointed appendages. The modeling of the ostracod carapace and appendages is so minute and complex that it is used in the classification of these animals. These distinguishing features are not visible under a standard light microscope, but are readily apparent when examined with a scanning electron microscope. The SEM's extraordinary depth of field at high magnification is also a great advantage in making microphotographs.

The preparation of the specimens used in this publication has been in collaboration with Dr. Louis Kornicker, research curator, Department of Invertebrate Zoology, National Museum of Natural History.

If ostracods are first narcotized, their valve systems will not completely retract. A mixture of chloral hydrate and menthol is recommended because this combination paralyzes the abductor muscles as it narcotizes the specimen.

A 2 percent solution of osmium tetroxide in distilled water is recommended for fixing. If the ostracod is of salt water origin, the fixing solution should be isotonically balanced to avoid distortion by the forces of osmotic pressure.

(Many specimens which are already part of functional collections have been fixed in ethanol and stored in a mixture of ethanol and glycerol. Good freeze-dry results are possible if the specimens are properly cleaned.)

After the specimens are fixed, they must be washed—with especial care—in several changes of filtered distilled water. Dr. Duane Hope of the Department of Invertebrate Zoology at the Smithsonian's National Museum of Natural History has designed a superb washing apparatus (figure 84).

To rid specimens of tenacious artifacts, put them in a vial of clean

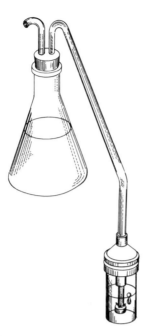

Figure 84. Hope Washing Apparatus.

distilled water, immersed in an ultrasonic cleaning bath. An ultrasonic pen cleaner is an ideal apparatus for such cleaning. Held with forceps, the open vial is suspended in the activated ultrasonic bath for about 10 sec.

The clean ostracod may be examined in a suspended drop on the underside of a cover slip, mounted over a concave glass slide. If examination with an oil immersion lens shows artifacts that obscure part of the specimen, more washing or ultrasonic cleaning is needed—perhaps by placing them in boiling nitrogen, then rapidly transferring them to clean distilled water. Finally, pipette them in single droplets to round cover slips that will fit the SEM stub.

After the cover slip is cooled from the underside with liquid nitrogen or a commercial spray freeze and the water droplets are frozen, it is transferred to a freeze-dry chamber. Any of these is suitable: Edward EF-2; Virtis Preservator; the cell apparatus illustrated in figure 18; or, with careful temperature control, the simple histology apparatus illustrated in figure 59. The specimen temperature should be maintained at −30° C. Drying time is 4 to 9 hrs. (overnight may be convenient and is not harmful).

After the specimens are dry, the chamber—while still under vacuum—is heated to room temperature, or the specimens are transferred to a desiccant jar. This protects the ostracod from rehydration from condensation on the cold cover slips.

METAL COATING

The cover slips containing the specimens are attached to the metal microscope stubs first with a drop of Duco cement, then with several pinpoint dots of metal-conductive paint, which assure electrical contact.

In the vacuum evaporator the assembly containing the specimen is coated with gold, gold-palladium, or aluminum. The coating must be uniform on all areas of the specimen. See figures 85-90.

Figure 85. *Anathron dethrix* Kornicker, 1975. Magnification: 88 diameters.

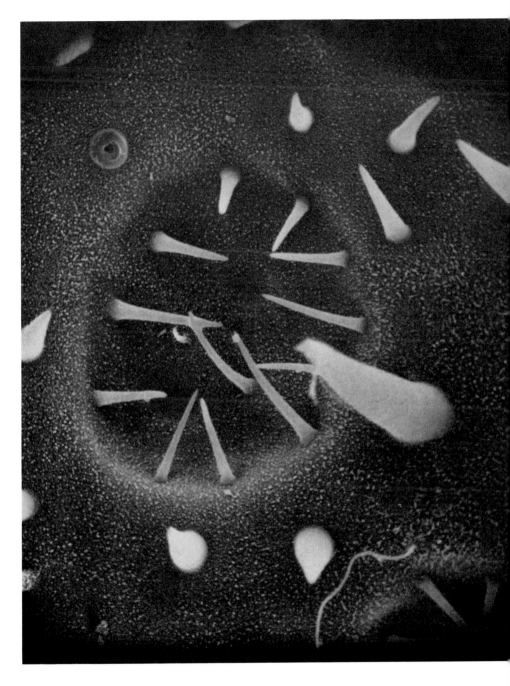

Figure 86. *Spinacopia bisetula* Kornicker, 1969. Details of bristles on outer carapace. Magnification: 2300 diameters.

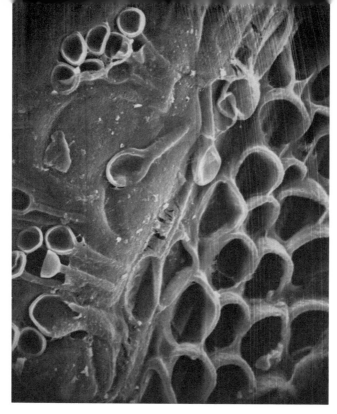

Figure 87. *Cymbicopia hanseni* (Brady, 1898). Bristles on outer carapace. This specimen was collected and preserved prior to its description in 1898. Magnification: 1600 diameters.

Figure 88. *Philomedes lofthousae* Kornicker, 1975. Tip of bristle on seventh limb. Magnification: 3500 diameters.

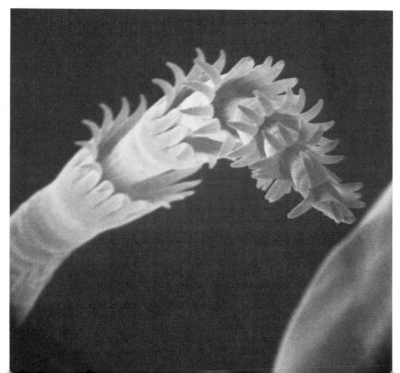

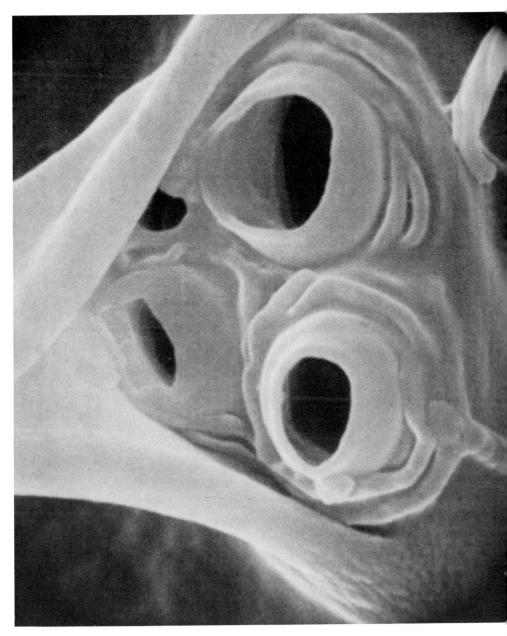

Figure 89. *Skogsbergiella spinifera* (Skogsberg, 1920). Pores, first revealed by SEM, are at base of flaplike bristle on posterior end of carapace. Magnification: 18000 diameters.

Figure 90. *Vargula subantarctica* Kornicker, 1975. Glandular openings on upper lip. Magnification: 2800 diameters.

NEMATODES

Nematodes, of the phylum Nemathelminthes, are round worms with unsegmented bodies that range in length from but a few micrometers to several centimeters.

In much the same manner as ostracods, nematodes may be fixed in a 2 percent solution of osmium tetroxide. Formalin fixation is also effective. Washing procedures are similar to those used with ostracods; in fact, the Hope washer was designed specifically for nematodes. Filtered distilled water flows over the specimens until they are washed clean (figure 84). See also figures 91-96.

Figure 91. Protozoa, *Suctoria*, on a marine nematode, *Desmodora*. Magnification: 1050 diameters.

Figure 92. Protozoa, *Suctoria*, on a marine nematode, *Desmodora*. Magnification: 2400 diameters.

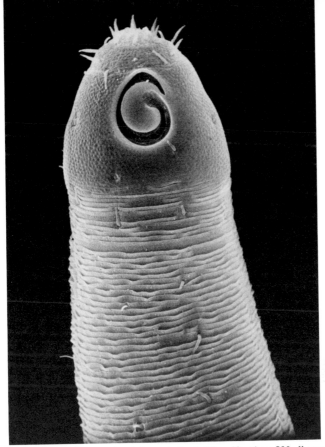

Figure 93. *Desmodora* head (lateral view). Magnification: 800 diameters.

Figure 94. *Enoplus* (face view). Magnification: 840 diameters.

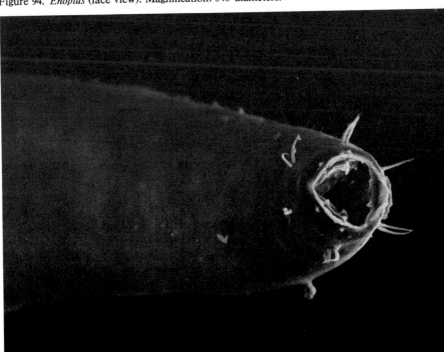

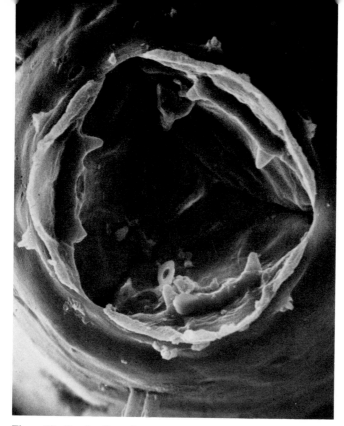

Figure 95. *Enoplus* (face view). Magnification: 2160 diameters.

Figure 96. *Eurystomina*. Open stoma. Magnification: 3600 diameters.

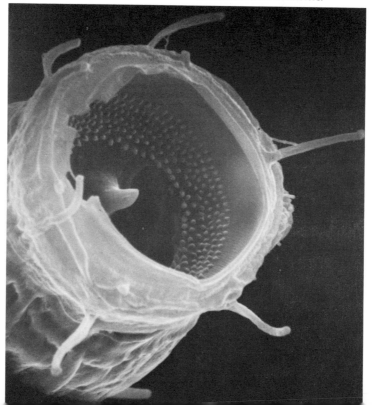

HUMAN BRAIN MATERIAL

Figures 97, 98, and 99 show samples of human brain which had been formalin-fixed and sliced, then washed and freeze-dried. When dry, a small specimen was removed at the line separating the cortical layer from the white medullary material, mounted on a SEM stub, and coated with gold.

Under the scanning electron microscope, the broken cell material showed a number of bodies (figure 97), each about 1.5 μm in diameter. These bodies have shown up in no brain area other than on the cortical boundary, but they have appeared in other cortical-boundary specimens prepared in different ways. They have been found in sheep brain and bovine brain. The fact that they are soluble in ethanol may explain why they have not been described: they would have disappeared during clearing.

Figure 97. Brain Body. Magnification: 43,000 diameters.

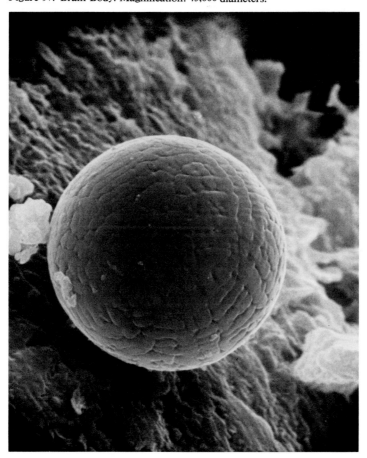

It has been suggested that these bodies may be concretions but this does not seem likely in view of their uniform spherical shape, their similarity in size and structure, and their quantity.

Figures 98 and 99 are scanning electron micrographs of exposed blood vessels. Figure 99 is a small capillary, about 10 μm across, also visible at the lower end of the larger vessel in figure 98.

At the Smithsonian we have experimented with varying degrees of success with bovine muscle, liver tissue, spines from *Acanthaster*, and many other types of biological material. There is much more work to be done.

Figure 98. Blood Vessels in Brain. Scanning electron micrograph. 700 diameters

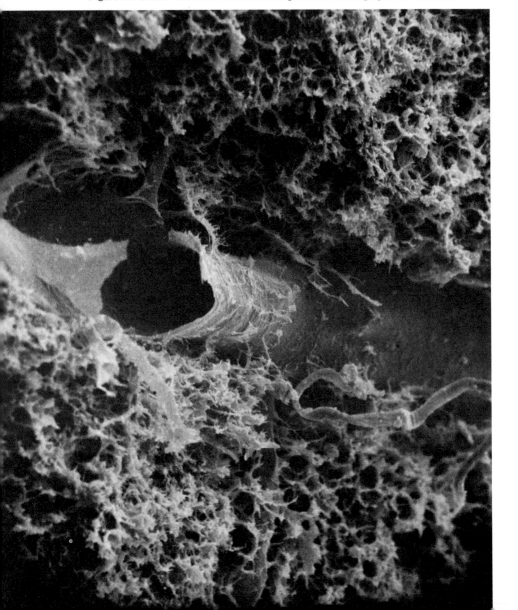

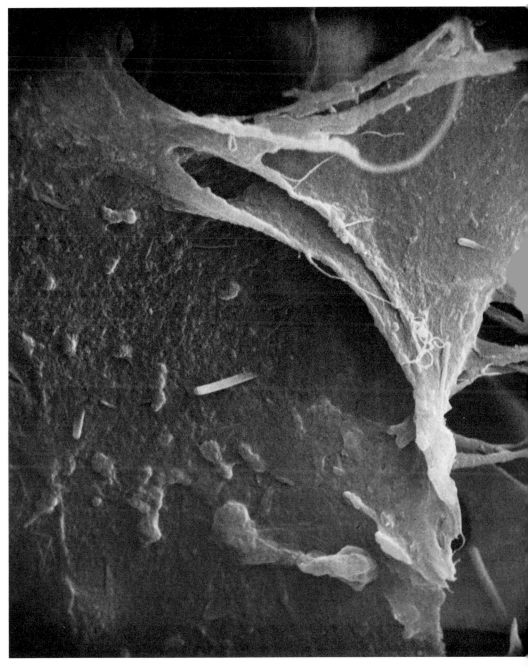

Figure 99. Capillary in Brain. Scanning electron micrograph. 70,000 diameters

PRESERVATION OF MARINE ARCHEOLOGICAL MATERIALS

A new application of the freeze-dry technique has evolved as a result of efforts by the National Park Service to retrieve the river steamer *Bertrand*, which sank in the Missouri River in 1865. Among the objects aboard was a shipment of merchants' almanacs. The almanac sent to the Smithsonian laboratory had been frozen in a commercial freezer—probably at about $-20°$ C—and was still frozen. Freshwater frozen at usual freezer temperatures (about $-20°$ C) forms large ice crystals which had developed between the pages of the almanac and caused them to separate.

Since the almanac was valuable, we experimented first with two sections from another old book (which was not valuable). The samples, which approximated the thickness and type of paper in the almanac, were watersoaked overnight, then frozen at $-20°$ C. Accurate weight records of each were kept throughout the experiment, and curves were plotted of the drying cycles.

Sample 1 was freeze-dried at $-20°$ C, until no further weight loss was evident. The pages were brittle and weighed less than before soaking. After a few hours in the laboratory atmosphere, they reabsorbed enough moisture to restore flexibility and weight.

Sample 2 was freeze-dried to the point that would closely approximate the post-drying weight recovery of sample 1.

Although, obviously, these experiments did not duplicate the original immersion conditions, the drying cycles were reasonably similar.

One section of the almanac was placed in the freeze-dry chamber at $-20°$ C. As with the samples, the chamber was then pumped down to a pressure of 150 μm Hg, but the almanac was removed before it was completely dry, slightly beyond the point chosen for sample 2.

After several hours in the laboratory atmosphere, the almanac's weight increase was almost imperceptible. The end result was a well-preserved specimen, with pages that could be separated with relative ease.

When the frozen almanac reached the laboratory, it weighed 952.9 g; final weight was 522.9 g. Total weight loss was 430 g or 45.125 percent (figure 100).

SALTED BEEF

Why the meat shipment that had been aboard the *Bertrand* was not reduced to a liquified mass stimulated the imagination as well as the olfactory organs of all who visited the Smithsonian's freeze-dry laboratory.

Even while frozen, the pungent smell of clostridial hydrolysis was evident in the 60-lb block of the beef. The fats had assumed a waxy quality and were partially saponified. The meat was fairly firm, but the outer surfaces were slightly crumbly.

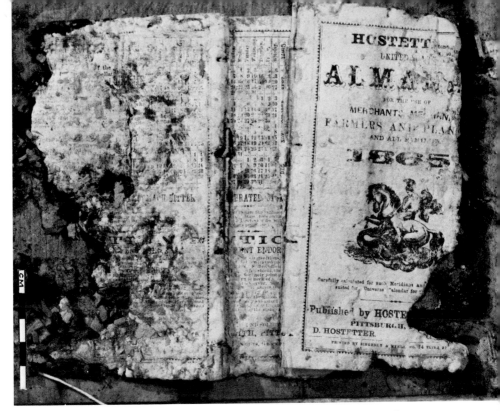

Figure 100. Almanac from the *Bertrand*. Dried at −20° C, 12 days (see drying chart, figure 145).

The drying process extended over a period of 12 wks (determined by weight loss) at a temperature of −20° C. The end result was a specimen that retained a slight putrid odor typical of clostridium, but was preserved sufficiently for study.

ELECTROLYTIC CONDUCTION DURING FREEZING

There appears to be a continuing effort on the part of researchers to reduce the time required to freeze-dry biological specimens.

Raising the temperature of the specimen surface will accomplish this, but at the expense of structural integrity. The price of shrinkage and distortion is too high in exchange for a small decrease in drying time.

An experiment aimed at reducing drying time was predicated on the following theory.

So far as is known, no organic chlorides exist in biological tissues. The pH of biological tissues apparently precludes any combination of protein-chloride. Since we know that chlorides—mainly potassium and sodium—are present in biological tissue, we may conclude that they are all inorganic. The chloride concentration in normal mammalian plasma is 5.6 to 6.3 g per l of sodium chloride (NaCl). Biological serum is approximately 9 g of NaCl per l.

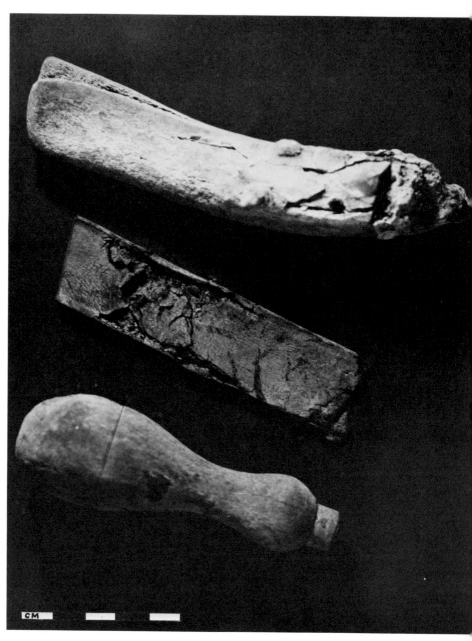

Figure 101. Items taken from blockade runners sunk during the Civil War. They were removed from marine archeological sites off the coast of North Carolina. Left to right: bone knife handle; fragment of a leather strap; wooden handle from a tool.

Figure 102. Fragment of Ship Timber. This material from a blockade runner that was sunk during the Civil War was collected from marine archeological sites off the coast of North Carolina.

Aqueous solutions of these salts are effective conductors of electricity. Consequently, the complete chemical structure of the biological system is an efficient conductor of electricity. Good conductors (or strong electrolytes) produce an abnormal decrease in the freezing point and a reduction in the vapor pressure (incidentally, an increase in the boiling point). Arrhenius, in his electrolytic dissociation theory, proposed that when substances that are good conductors of electricity are dissolved in water they are dissociated. Sodium chloride becomes: $NaCl \rightleftharpoons Na^+ + Cl^-$ chlorine and sodium ions, each electrically charged. In terms of the effect upon the boiling and freezing points, each ion behaves as a unit, so that when NaCl is dissolved it acts like two ions ($Na^+ + Cl^-$) rather than one molecule. As a concentrate solution of NaCl is diluted, the effect is an electrolytic dissociation of existing ions, not an ionization of preexisting molecules. Therefore, the electrostatic attraction is greater between ions with higher charges. This explains the difference between NaCl, for example, and $MgSO_4$. The doubly charged (Mg^{++}) and sulfate (SO_4^{--}) have greater attraction for each other than (Na^+) and (Cl^-) and consequently have less effect on depressing the freezing point.

When two electrodes are immersed in a normal saline electrolyte and are oppositely charged, the positive ions will be attracted to the negative electrode or cathode and the negative ions will be attracted to the positive electrode or anode.

Since there is no net gain or loss of electrons, the number of electrons accepted by the Na^+ ions is equal to the number of electrons given by the Cl^- ions. The electrolytic reactions are: $H_2O \rightleftharpoons H^+ + OH^-$ at the anode $2Cl^- \rightarrow 2e^- + Cl_2\uparrow$ at the cathode $2H_2O + 2e \rightarrow H_2 + 2OH^-$ Na^+ ions attracted to the cathode, $+ OH^-$ formed there $Na^+ + OH^- \rightleftharpoons NaOH$.

In an effort to determine the pattern of ion transfer, bovine muscle tissues were cut to uniform size, and electrodes were installed. An electric current of 10 ma was passed through the samples at 100 v. The movement of charged particles through the solution results in the transfer of electrical charge, i.e., electric current.

This specific conduction, which results from the movement of ions through the solution, is called ionic or electrolytic conduction. The amount of the charge moved through the solution in a given time is dependent upon the number of ions present and the speed at which they move, which is, in turn, dependent to a small degree upon whether the ion is monovalent, divalent, or polyvalent.

Numbers, degree, and speed notwithstanding, electrolysis is the mechanism of ion transport through a solution, as the result of electrolytic conduction. Sodium chloride solutions, whether or not they are associated with biological tissue, behave in the same manner.

The bovine-tissue specimens were partially dried for exactly the same period of time and then cut and photographed. Figure 103 is a typical control sample frozen at $-70°$ C, with electrolysis.

Figure 103. Bovine Tissue (control), partially dried to demonstrate normal freeze boundaries.

Figure 104. Bovine Tissue, partially dried to demonstrate the effect of electrolytic conduction during freezing.

Conclusion

In the presence of the electrodes there was a definite change in freezing pattern. Eutectic formation was established in a straight line between the anode and the cathode. Also, there was deposition of salts in the vicinity of the cathode.

The electrodes introduced a significant variation in specific heat (figure 105).

Figure 105. Freezing Variations During Electrolytic Conduction.

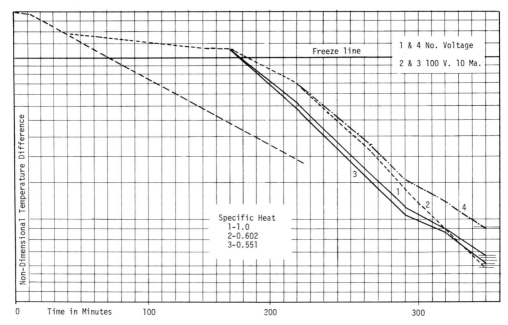

APPENDIX A

Drying-Rate Charts

If a freeze-dry system is loaded with many specimens releasing water vapor at a high rate, and the condenser is not capable of handling the vapor load, decided differences will occur in the drying time of each specimen. The following drying charts are useful, with recognition of the demands and limitations of the system at any given time. Nearly dry specimens may actually gain weight; others may show change in drying rate. Since most systems are subject to such variations, the charts are based on averages.

When drying temperature is reduced, drying time is radically increased. Temperature reduction is frequently a good compromise to maintain histological integrity of specimens that might be adversely affected by small amounts of shrinkage.

The bovine-muscle charts do not reflect system limitations but demonstrate the effects of freezing temperatures on drying cycles. Some samples frozen at lower temperatures demonstrated a uniform but slower drying cycle, while other specimens, frozen at higher temperatures, were partially dry in less time. The total weight loss, however, was reduced over the three-day drying cycle and the specimens frozen at $-30°$ C contained up to 20 percent more water than the specimens frozen at $-70°$ C. This water retention was due to the concentration of salts during freezing—which supports the theory that much of the success in the freeze-dry process is dependent upon rapid freezing. Prechilling the specimen to a point just above freezing also demonstrated an apparent advantage, as shown in figures 109 and 112. The weight gain in samples 1 and 2 was due to the concentration of inorganic salts that absorbed water of crystalization, and resulted in ice with a lower vapor pressure.

Figure 106 demonstrates several conditions as they apply to drying rates. The surface-to-mass ratio is calculated as a spherical model in the following equation:

$$r_a = \left[\frac{3}{4\pi}\left(\frac{Ma\text{-}Ms}{1\text{-}s}\right)\right]1/3 \text{ and } S = r_a^2 4\pi$$

where: M_a = mass of solid + H_2O at any time
M_s = mass of solid (dried tissue)
ρ_s = apparent density of the solid
r_a = radius of the ice mass at the drying front
s_a = surface area at the drying front

For the purpose of this figure, all dimensions were calculated to a unit of 1 with the equation:

$$\frac{Ma - Ms}{Ma_o - Ms} = Ma{:}1, \frac{Sa}{Sa_o} = S_a{:}1$$

$$\frac{\triangle L}{\triangle L_f} = \triangle{:}1 \text{ and } \frac{t}{t_f} = t{:}1$$

where: Ma_o = the original mass of the solid and H_2O
$\quad\quad Sa_o$ = the original surface area
$\quad\quad \triangle L$ = the thickness of the dried layer
$\quad\quad \triangle L_f$ = the radius of the complete dried mass
$\quad\quad t_f$ = the total drying time of the ice mass
for ice $\rho_s = i$, hence the equation:

$$r_a = \left[\frac{3}{4\pi}(M_a - M_s)\right]^{1/3}$$

Note that with a surface-to-mass ratio of 1, the lines plotted for ice and for the ice mass within the specimen where $\rho_s = .2$, the lines are nearly superimposed.

The variation in the surface-to-mass relationship compared with the mass-to-time relationship varies with the thickness of the dried tissue layer. The impedance offered by the dried layer bears an almost linear relationship to drying time.

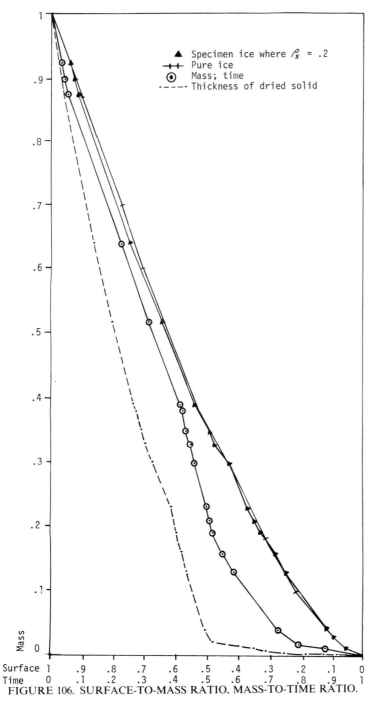

FIGURE 106. SURFACE-TO-MASS RATIO, MASS-TO-TIME RATIO.

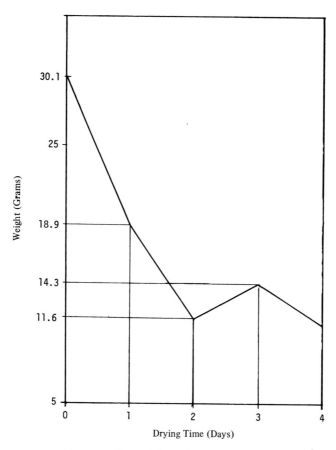

Bovine muscle sample frozen from room temperature to -30° C., dried at -30° C.
Initial weight: 30.1 g.
Sample #1

FIGURE 107. BOVINE MUSCLE DRYING SAMPLE #1.

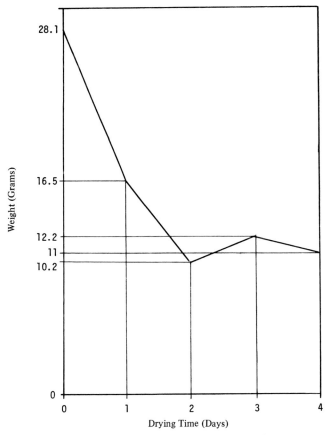

Bovine muscle sample frozen from 37° C. to -30° C., dried at -30° C.
Initial weignt: 28.1 g.
Sample #2

FIGURE 108. BOVINE MUSCLE DRYING SAMPLE #2.

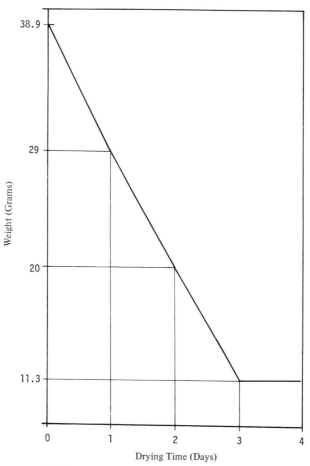

Bovine muscle samples frozen from 0° C. to -30° C., dried at -30° C.
Initial weight: 38.9 g.
Sample #3

FIGURE 109. BOVINE MUSCLE DRYING SAMPLE #3.

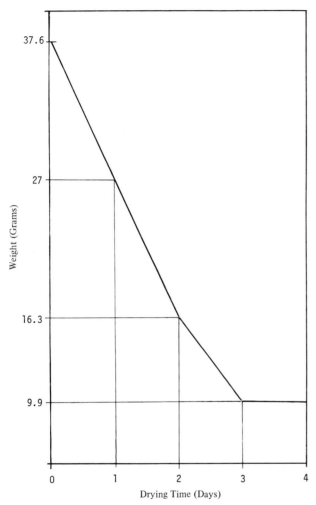

Bovine muscle sample frozen from room temperature to -70° C., dried at -30° C.
Initial weight: 37.6 g.
Sample #4

FIGURE 110. BOVINE MUSCLE DRYING SAMPLE #4.

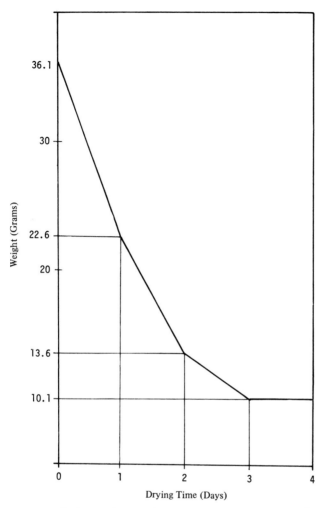

Bovine muscle sample frozen from room temperature to -70° C., dried at -30° C.
Initial weight: 36.1 g.
Sample #5

FIGURE 111. BOVINE MUSCLE DRYING SAMPLE #5.

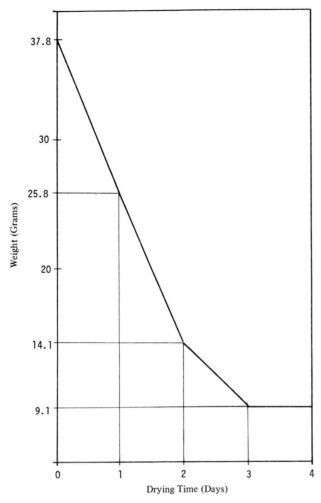

Bovine muscle sample frozen from 0° C. to -70° C., dried at -30° C.
Initial weight: 37.8 g.
Sample #6

FIGURE 112. BOVINE MUSCLE DRYING SAMPLE #6.

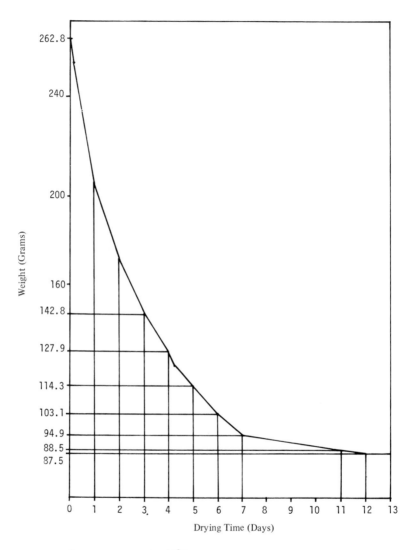

Baboon lungs, dried at -30°C.
Initial weight: 262.8 g.

FIGURE 113. BABOON LUNGS DRYING RATE.

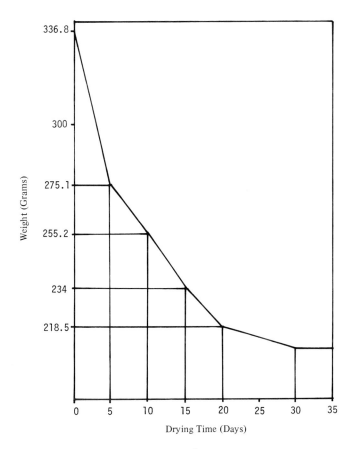

Human stomach, dried at -25° C.
Initial weight: 336.8 g.

FIGURE 114. HUMAN STOMACH DRYING RATE.

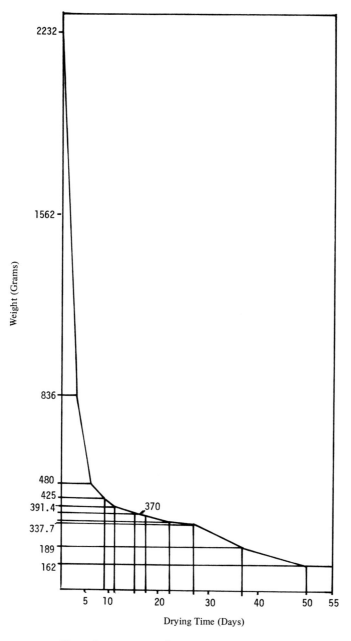

Human fetus, dried at -30° C.
Initial weight: 2232 g.

FIGURE 115. HUMAN FETUS DRYING RATE.

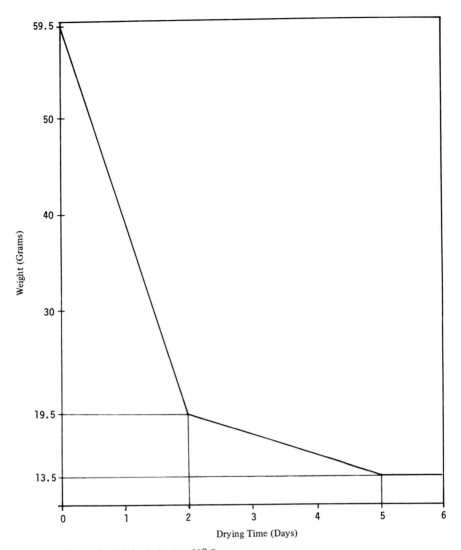

Human thyroid gland, dried at -30° C.

Initial weight: 59.5 g.

FIGURE 116. HUMAN THYROID DRYING RATE.

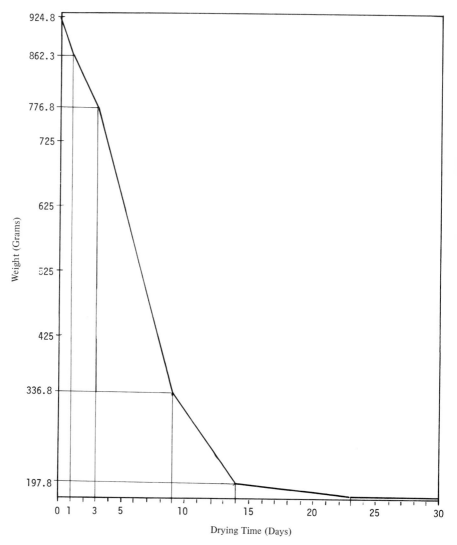

Human heart, coronary arteriosclerosis, dried at -30° C.
Initial weight: 924.8 g.

FIGURE 117. HUMAN HEART DRYING RATE.

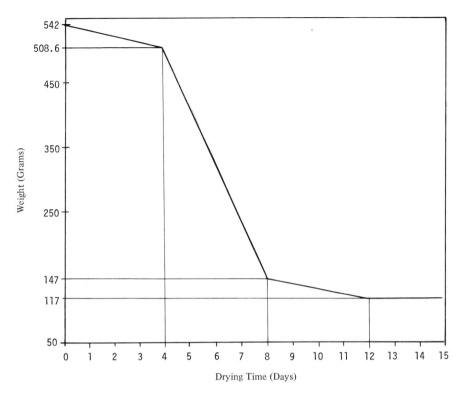

Human cardiac material, dried at -30° C.
Initial weight: 542.0 g.

FIGURE 118. HUMAN CARDIAC MATERIAL DRYING RATE.

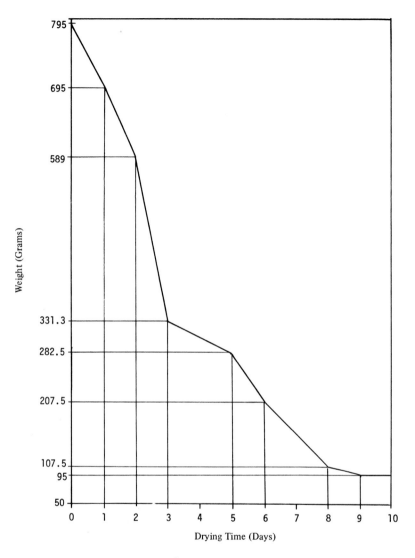

Human lung, dried at -30° C.

Initial weight: 795.0 g.

FIGURE 119. HUMAN LUNG DRYING RATE.

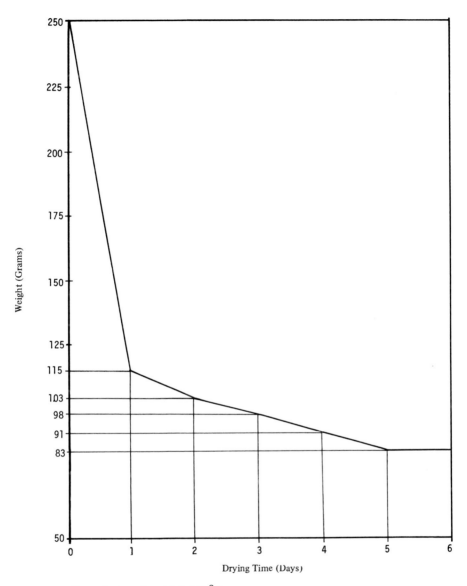

Human brain section, dried at -30° C.

Initial weight: 250.0 g.

FIGURE 120. HUMAN BRAIN SECTION DRYING RATE.

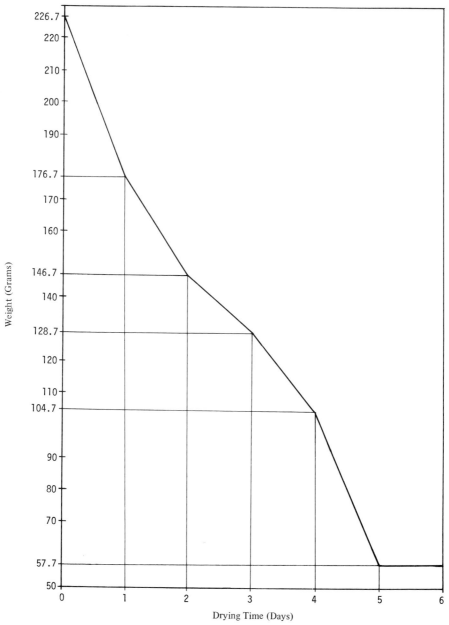

Human brain section, dried at -30° C.
Initial weight: 226.7 g.

FIGURE 121. HUMAN BRAIN SECTION DRYING RATE.

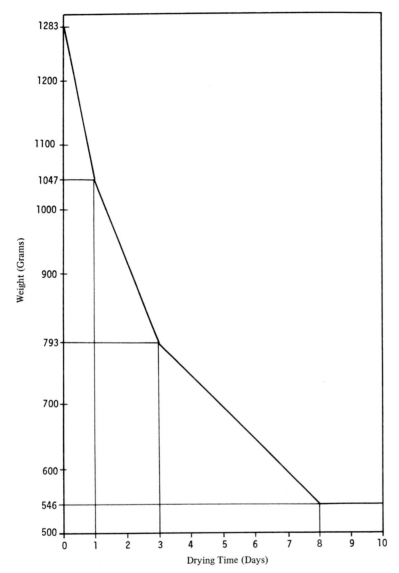

Sectioned human brain, dried at -30° C.
Initial weight: 1283 g.

FIGURE 122. HUMAN BRAIN SECTION DRYING RATE.

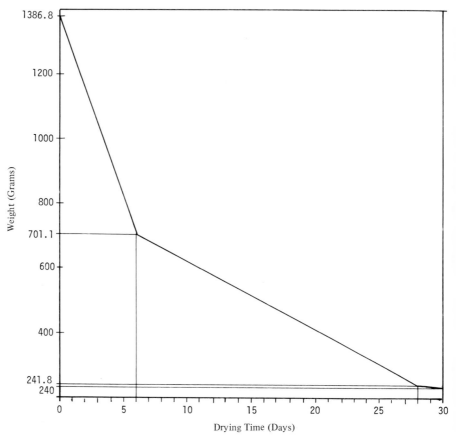

Complete human brain, dried at -30° C.
Initial weight: 1386.8 g.

FIGURE 123. BRAIN, COMPLETE DRYING RATE.

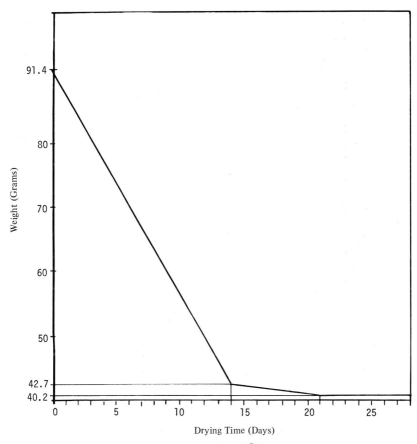

Eastern Chipmunk, *Tamias striatus,* dried at -30° C.
Initial weight: 91.4 g.

FIGURE 124. EASTERN CHIPMUNK DRYING RATE.

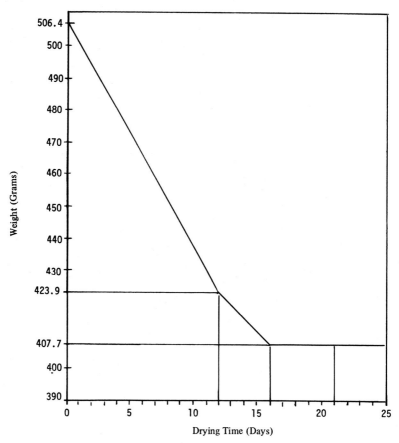

Bats Group #1, *Eptesicus fuscus,* dried at -20° C.
Initial weight: 506.4 g.

FIGURE 125. CAVE BATS DRYING RATE.

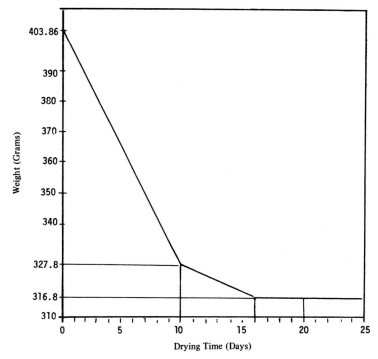

Bats Group #2, *Eptesicus fuscus,* dried at -20° C.
Initial weight: 403.86 g.

FIGURE 126. CAVE BATS DRYING RATE.

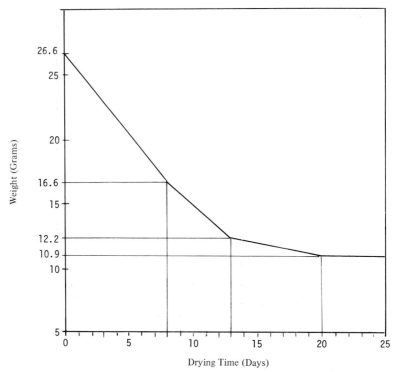

Meadow Vole, dried at -20° C.
Initial weight: 26.6 g.

FIGURE 127. MEADOW VOLE DRYING RATE.

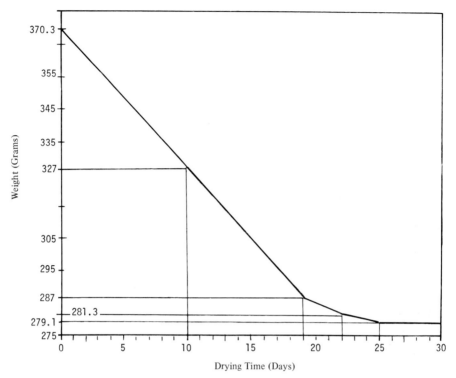

Mourning Dove, *Zeniadura macroura,* dried at -20° C.
Initial weight: 370.3 g.

FIGURE 128. MOURNING DOVE DRYING RATE.

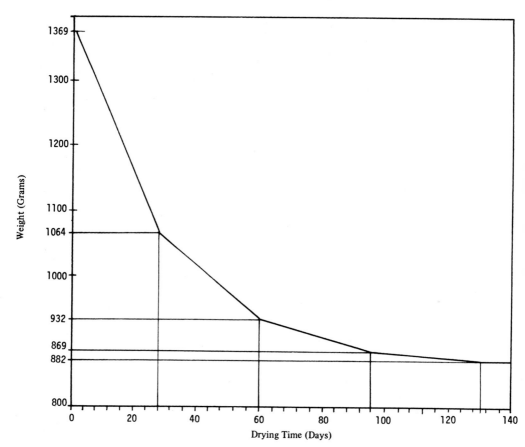

Barred Owl, *Strix varia*, dried at -20° C.
Initial weight: 1369 g.

FIGURE 129. BARRED OWL DRYING RATE.

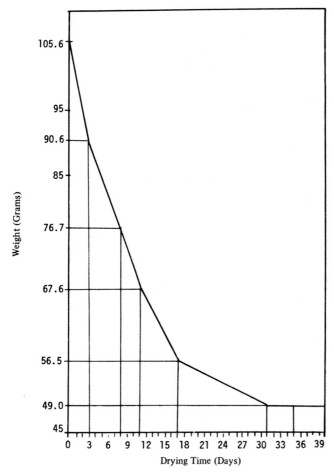

Red-Bellied Woodpecker, *Centurus carolinus,* dried at -25° C.

Initial weight: 105.6 g.

FIGURE 130. RED-BELLIED WOODPECKER DRYING RATE.

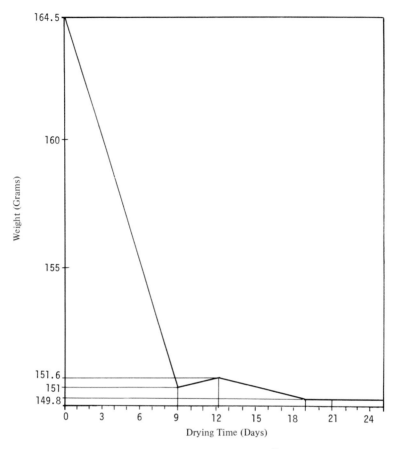

Slate-Colored Junco, *Junco hyermalis,* dried at -20° C.
Initial weight: 164.5 g.

FIGURE 131. SLATE-COLORED JUNCO DRYING RATE.

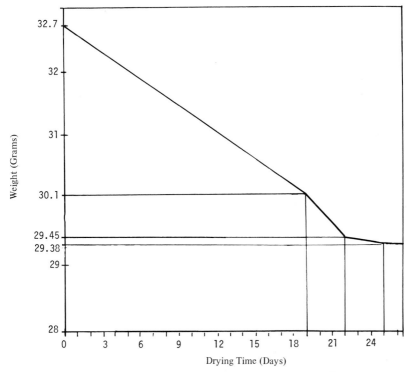

Ruby-Throated Hummingbird, *Archilochus colubris,* dried at -30° C.
Initial weight: 32.7 g.

FIGURE 132. RUBY-THROATED HUMMINGBIRD DRYING RATE.

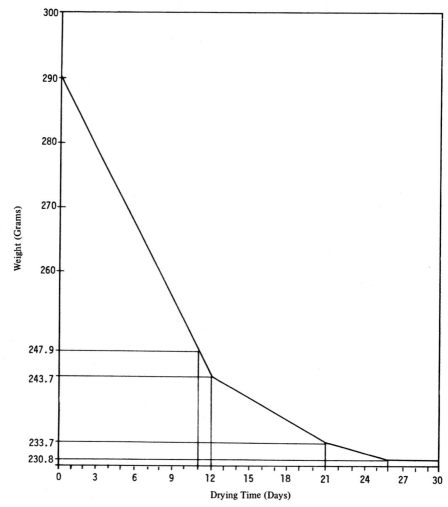

Virginia Rail, *Rallus limicola limicola,* dried at -20° C.
Initial weight: 290.4 g.

FIGURE 133. VIRGINIA RAIL DRYING RATE.

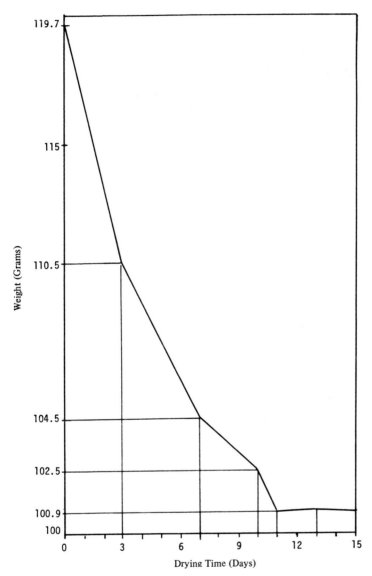

Eastern Chipping Sparrow, *Spizella passerina passerina*, dried at -20° C.
Initial weight: 119.7 g.

FIGURE 134. EASTERN CHIPPING SPARROW DRYING RATE.

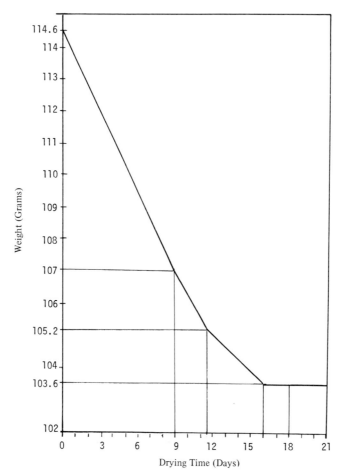

Hooded Warbler, *Wilsonia citrina,* dried at -20° C.
Initial weight: 114.6 g.

FIGURE 135. HOODED WARBLER DRYING RATE.

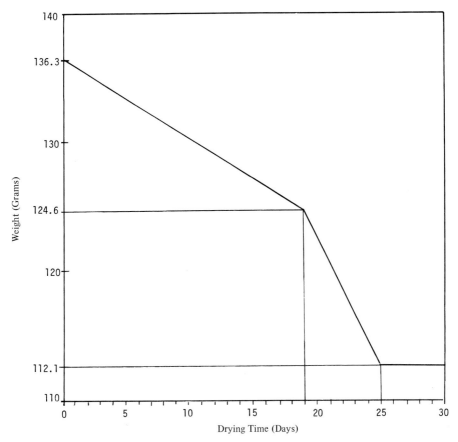

Philadelphia Verio, *Verio philadelphicus,* dried at -30° C.
Initial weight: 136.3 g.

FIGURE 136. PHILADELPHIA VERIO DRYING RATE.

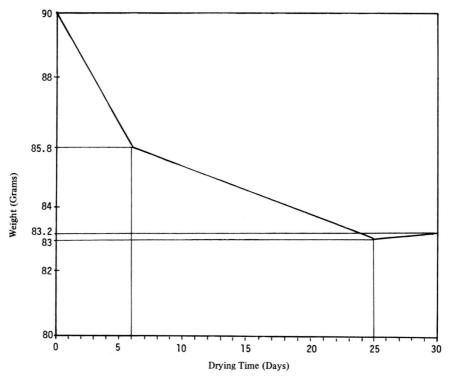

Brown Creeper, *Certhia familiaris,* dried at -30° C.
Initial weight: 90 g.

FIGURE 137. BROWN CREEPER DRYING RATE.

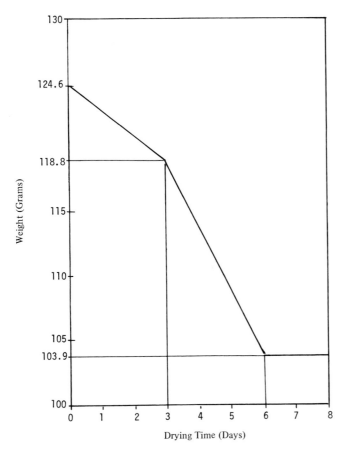

Cedar Waxwing, *Bombycilla cedrorum,* dried at -15° C.
Initial weight: 124.6 g.

FIGURE 138. CEDAR WAXWING DRYING RATE.

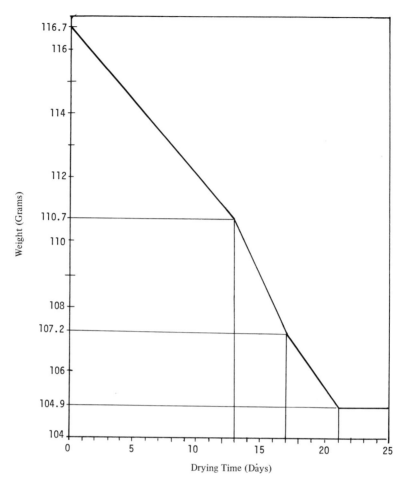

Indigo Bunting, *Passerina cyanea,* dried at -20° C.
Initial weight: 116.7 g.

FIGURE 139. INDIGO BUNTING DRYING RATE.

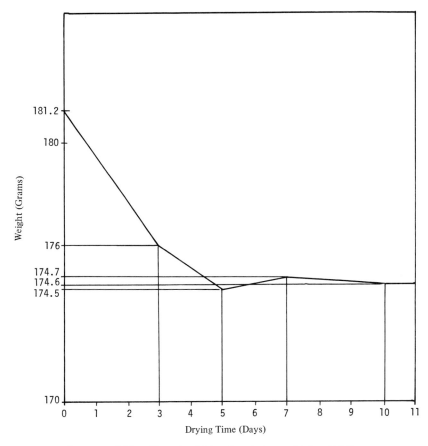

Cottonmouth Water Moccasin, *Agkistrodon piscivorus,* dried at -30° C.
Initial weight (eviscerated): 181.2 g.

FIGURE 140. COTTONMOUTH WATER MOCCASIN DRYING RATE.

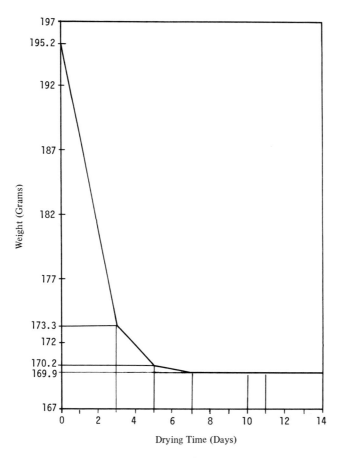

Green Water Snake, dried at -30° C.

Initial weight (eviscerated): 195.2 g.

FIGURE 141. GREEN WATER SNAKE DRYING RATE.

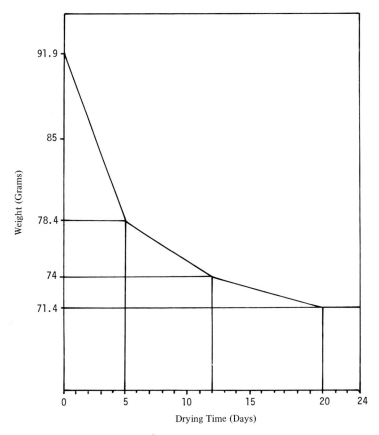

Octopus, dried at -25° C.

Initial weight: 91.9 g.

FIGURE 142. OCTOPUS DRYING RATE.

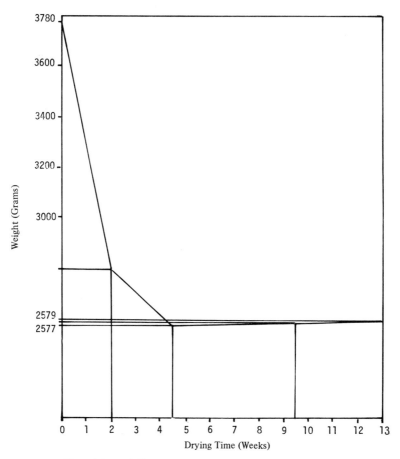

Skate, dried at -25° C.
Initial weight: 3780 g.

FIGURE 143. SKATE DRYING RATE.

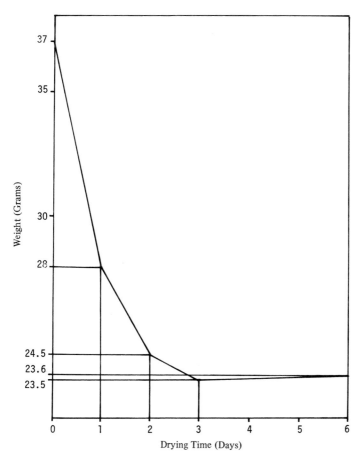

Hermit Crab, dried at -20° C.
Initial weight: 37 g.

FIGURE 144. HERMIT CRAB DRYING RATE.

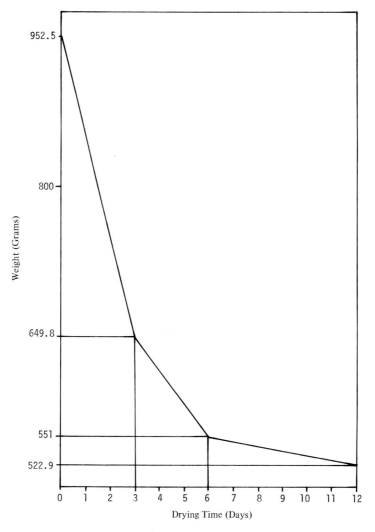

Almanac, dried at -20° C.

Initial weight: 952. 5 g.

FIGURE 145. ALMANAC DRYING RATE.

APPENDIX B

TABLE 12.—TEMPERATURE CONVERSION

The numbers in the reference column represent the temperature in either degrees Celsius or degrees Fahrenheit. For easy conversion from degrees Celsius, the Fahrenheit equivalent will be found in the right-hand column. To convert from degrees Fahrenheit, the Celsius equivalent is found in the left-hand column.

C	Reference	F	C	Reference	F	C	Reference	F
−273	−459.4		−129	−200	−328	−13.9	7	44.6
−268	−450		−123	−190	−310	−13.3	8	46.4
−262	−440		−118	−180	−292	−12.8	9	48.2
−257	−430		−112	−170	−274	−12.2	10	50
−251	−420		−107	−160	−256	−11.7	11	51.8
−246	−410		−101	−150	−238	−11.1	12	53.6
−240	−400		−95.6	−140	−220	−10.6	13	55.4
−234	−390		−90	−130	−202	−10	14	57.2
−229	−380		−84.4	−120	−184	−9.44	15	59
−223	−370		−78.9	−110	−166	−8.89	16	60.8
−218	−360		−73.3	−100	−148	−8.33	17	62.6
−212	−350		−67.8	−90	−130	−7.78	18	64.4
−207	−340		−62.2	−80	−112	−7.22	19	66.2
−201	−330		−56.7	−70	−94	−6.67	20	68
−196	−320		−51.1	−60	−76	−6.11	21	69.8
−190	−310		−45.6	−50	−58	−5.56	22	71.6
−184	−300		−40	−40	−40	−5	23	73.4
−179	−290		−34.4	−30	−22	−4.44	24	75.2
−173	−280		−28.9	−20	−4	−3.89	25	77
−169	−273	−459.4	−23.3	−10	14	−3.33	26	78.8
−168	−270	−454	−17.8	0	32	−2.78	27	80.6
−162	−260	−436	−17.2	1	33.8	−2.22	28	82.4
−157	−250	−418	−16.7	2	35.6	−1.67	29	84.2
−151	−240	−400	−16.1	3	37.4	−1.11	30	86
−146	−230	−382	−15.6	4	39.2	−.56	31	87.8
−140	−220	−364	−15	5	41	−0.0	32	89.6
−134	−210	−346	−14.5	6	42.8			

TABLE 13.—UNDERWRITERS' LABORATORIES CLASSIFICATION OF COMPARATIVE LIFE HAZARDS OF GASES AND VAPORS

Group	Definition	Examples
1	Gases or vapors which in concentrations of the order of $\frac{1}{2}$ to 1% for durations of exposure of the order of five minutes are lethal or cause serious injury.	Sulfur dioxide
2	Gases or vapors which in concentrations of the order of $\frac{1}{2}$ to 1% for durations of exposure of the order of $\frac{1}{2}$ hour are lethal or produce serious injury.	Ammonia methyl bromide
3	Gases or vapors which in concentrations of the order of 2 to 2% for durations of exposure of the order of 1 hour are lethal or produce serious injury.	Bromochloromethane Carbon tetrachloride Chloroform Methyl formate
4	Gases or vapors which in concentrations of the order of 2 to $2\frac{1}{2}$% for durations of exposure of the order of 2 hours are lethal or cause serious injury.	Dichloroethylene Methyl chloride Ethyl bromide
Between 4&5	Appear to classify as somewhat less toxic than group 4. Much less toxic than group 4 but somewhat more toxic than group 5.	Methylene chloride Ethyl chloride Refrigerant 112 Refrigerant 113* Refrigerant 21
5a	Gases or vapors much less toxic than group 4 but more toxic than group 6.	Refrigerant 11 Refrigerant 22 Refrigerant 114 B2 Refrigerant 502 Carbon dioxide
5b	Gases or vapors which available data would classify as either group 5a or group 6.	Ethane propane Butane
6	Gases or vapors which in concentrations up to at least about 20% by volume for durations of exposure of the 2 hours do not appear to produce injury.	Refrigerant 13B1 Refrigerant 12 Refrigerant 114 Refrigerant 115 Refrigerant 13* Refrigerant 14* Refrigerant 23* Refrigerant 116* Refrigerant C318*

*Not tested by UL but estimated to belong in group indicated.

TABLE 14.—TEMPERATURE-PRESSURE TABLE FOR VARIOUS REFRIGERANTS

Temperature		Refrigerants bold face—inches of mercury; roman—psi			
F°	C°	R.12	R.22	R.502	R.717
−100	−73.3	**27.0**	**25.0**	**23.3**	**27.4**
−95	−70.6	**26.4**	**24.0**	**22.1**	**26.8**
−90	−67.8	**25.7**	**32.0**	**20.6**	**26.1**
−85	−65.0	**25.0**	**21.7**	**19.0**	**25.3**
−80	−62.2	**24.0**	**20.2**	**17.1**	**24.3**
−75	−59.4	**23.0**	**18.5**	**15.0**	**23.2**
−70	−56.7	**21.8**	**16.6**	**12.6**	**21.9**
−65	−53.9	**20.5**	**14.4**	**10.0**	**20.4**
−60	−51.5	**19.0**	**12.0**	**7.0**	**18.6**
−55	−48.3	**17.3**	**9.2**	**3.6**	**16.6**
−50	−45.6	**15.4**	**6.2**	**0.0**	**14.3**
−45	−42.8	**13.3**	**2.7**	2.1	**11.7**
−40	−40.0	**11.0**	**0.5**	4.3	**8.7**
−38	−38.9	**10.0**	1.3	5.2	**7.4**
−36	−37.7	**8.9**	2.2	6.2	**6.1**
−34	−36.7	**7.8**	3.0	7.2	**4.7**
−32	−35.6	**6.7**	3.9	8.3	**3.2**
−28	−33.3	**4.3**	5.8	10.5	**0.0**
−26	−32.2	**3.0**	6.9	11.7	0.8
−24	−31.1	**1.6**	7.9	13.0	1.7
−22	−30.0	**0.3**	9.0	14.2	2.6
−20	−28.9	0.6	10.1	15.5	3.6
−18	−27.8	1.3	11.3	16.9	4.6
−16	−26.7	2.0	12.5	18.3	5.6
−14	−25.6	2.8	13.8	19.9	6.7
−12	−24.4	3.6	15.1	21.2	7.9
−10	−23.3	4.5	16.5	22.8	9.0
−8	−22.2	5.4	17.9	24.4	10.3
−6	−21.1	6.3	19.3	26.0	11.6
−4	−20.0	7.2	20.8	27.7	12.9
−2	−18.9	8.2	22.4	29.4	14.3
0	−17.8	9.2	24.0	31.2	15.7
1	−17.2	9.7	24.8	32.2	16.5
2	−16.7	10.2	25.6	33.1	17.2
3	−16.1	10.7	26.4	34.1	18.0
4	−15.6	11.2	27.3	35.0	18.8
5	−15.0	11.8	28.2	36.0	19.6
6	−14.4	12.3	29.1	37.0	20.4
7	−13.9	12.9	30.0	38.0	21.2
8	−13.3	13.5	30.9	39.0	22.1
9	−12.8	14.0	31.8	40.0	22.9
10	−12.2	14.6	32.8	41.1	23.8
11	−11.7	15.2	33.7	42.2	24.7
12	−11.1	15.8	34.7	43.2	25.6
13	−10.6	16.4	35.7	44.3	26.5
14	−10.0	17.1	36.7	45.4	27.5

TABLE 14.—TEMPERATURE-PRESSURE TABLE FOR VARIOUS REFRIGERANTS (cont.)

Temperature		Refrigerants bold face—inches of mercury; roman—psi			
F°	C°	R.12	R.22	R.502	R.717
15	− 9.4	17.7	37.7	46.6	28.4
16	− 8.9	18.4	38.7	47.7	29.4
17	− 8.3	19.0	39.8	48.9	30.4
18	− 7.8	19.7	40.8	50.1	31.4
19	− 7.2	20.4	41.9	51.2	32.5
20	− 6.7	21.0	43.0	52.4	33.5
21	− 6.1	21.7	44.1	53.7	34.6
22	− 5.6	22.4	45.3	54.9	35.7
23	− 5.0	23.2	46.0	56.2	36.8
24	− 4.4	23.9	47.6	57.4	37.9
25	− 3.9	24.6	48.8	58.7	39.0
26	− 3.3	25.4	49.9	60.0	40.2
27	− 2.8	26.1	51.2	61.4	41.4
28	− 2.2	26.9	52.4	62.7	42.6
29	− 1.7	27.7	53.6	64.1	43.8
30	− 1.1	28.4	54.9	65.4	45.0
31	− 0.6	29.2	56.2	66.8	46.3
32	0	30.1	57.5	68.2	47.6

GLOSSARY

Absolute zero	temperature (equal to $-273.159°$ C ($-459.69°$ F)) at which a gas would have no pressure, assuming that the general gas law holds true at all temperatures.
Acid	substance that gives up a proton. Compounds that dissociate in water solution to produce H_3O^+ or H^+ (aq) ions are called acids; pH below 7.
Anion	negatively charged ion.
Anode	positively charged electrode in an electrolytic cell; the electrode at which oxidation occurs; electrode toward which anions travel.
Avagadro's Law	equal volumes of different gasses at the same temperature and at equal pressures contain the same number of molecules.
Azeotropic mixture	mixture of two or more liquids that behaves like a single substance, i.e., vapors from azeotropic mixtures produced by partial evaporation have the same composition as the liquid mixture.
Bar	international pressure unit equal to 10^6 dynes/cm^2 or 0.987 atmosphere.
Barye	international pressure unit equal to 1 dyne/cm^2.
Base	substance that dissociates on solution in water producing one or more hydroxyl ions; pH measurement above 7.
Boyle's Law	with temperature constant, the volume (v) of a given quantity of gas varies inversely with pressure (P_2); $Pv = pv$.
Btu	British thermal unit, amount of heat required to raise the temperature of 1 lb of water $1°$ F at $39.1°$ F. One BTU is equivalent to 0.252 kg calorie.

Calorie	amount of heat required to raise the temperature of 1 g of water 1° C at 15° C. The National Bureau of Standards defines a calorie as 4.184 joules; the international calorie (London 1929) is defined as 1/860 of the international watt hour or 4.1860 joules.
Cathode	negatively charged electrode in an electrolytic cell; electrode at which reduction occurs; electrode toward which cations travel.
Cation	positively charged ion.
Centipoise	standard unit of viscosity equal to .01 poise (cgs unit of viscosity).
Cgs	centimeter gram second.
Charles' (Gay-Lussac's) Law	volumes assumed by a given mass of gas (at a constant pressure) at different temperatures are directly proportional to the corresponding absolute temperatures.
Concentration	amount of a substance contained in a given volume of solution, stated by weight, moles, or equivalents.
Conductivity (thermal)	time rate of heat transfer by conduction through a given change in temperature; calories per sec per cm and temperature difference of 1° C. If two opposing surfaces of a solid are maintained at temperatures t_1 and t_2, the heat conducted across the solid of section a and the thickness d in time T are $$K = \frac{t_1 - t_2 aT}{d}$$ K is the constant dependent upon the nature of the substance and is usually given for Q in calories, a in cm^2, T in seconds, t in C° and d in cm.
Conductance (electrolytic)	reciprocal of resistance, the ratio of current flowing through a conductor to the difference in potential between its ends, or the quantity of electricity transferred across a unit area per unit potential in unit time.
Density	concentration of matter measured by mass per unit volume.

Dyne	force required to accelerate 1 g of mass 1 cm per second.
Electrolyte	substance capable of existing in solution as ions when placed in water or other dissociating solvents. A part of the substance passes into solution in the form of ions, leaving the remaining portion charged with an equivalent but opposite charge of electricity from that transferred by the ion.
Electrolysis	current i flows for time t and deposits a substance whose electrochemical equivalent is e. The deposited substance has mass m. Where $m = eit$ e = mass in grams, i = current in amperes and t = time in seconds.
Enthalpy	heat constant that is a thermodynamic quantity equal to the sum of the internal energy of a system plus the product of the pressure volume work done on the system. $H = E + pv$ (where H = enthalpy or heat constant; E = internal energy of the system; and p = pressure; v = volume).
Entropy	mathematical capacity factor for energy that is isothermally unavailable in a thermodynamic system. The increase in the entropy of a body during an infinitesimal stage of a reversible process is equal to the amount of absorbed heat divided by the absolute temperature of the body ($dS = Q/T$). All spontaneous natural processes demonstrate an increase in total entropy of the bodies involved in the process.
Eutectic	a solution having the lowest possible melting point; as used here, mixture of substances resulting in a depressed melting/freezing point. In biological fluids, depression of freezing point is due chiefly to presence of inorganic salts and organic esters.
Fluidity	reciprocal of viscosity, reciprocal of poise.
Gas	state of matter wherein molecules are practically unrestricted by cohesive forces, retaining neither specific shape nor volume.

Gay-Lussac's Law	see Charles' Law.
Gram mole	numerical equality of mass in grams to the molecular weight.
Heat capacity	amount of heat usually expressed in calories, required to increase the temperature of a substance 1° C.
Hg	mercury; used here to denote measurement of barometric pressure.
Humidity (absolute)	amount of water vapor in a given volume of air stated in grams per cubic meter.
Isotherm	gas undergoing changes in volume or pressure at a constant temperature; graphic line denoting the constant temperature.
Joule-Thompson effect	cooling that occurs when a highly compressed gas is expanded in such a way that no work is done. Such cooling is inversely proportional to the square of the absolute temperature.
Kinetic theory of gases	minute elastic particles of a gas move about at high velocities, colliding with one another and with the walls of their container, creating pressure.
Latent heat of fusion	amount of heat required to change 1 g of solid to a liquid with no change in temperature. For ice the latent heat of fusion is 80 calories.
Latent heat of sublimation	as used here, applies to the sublimation of ice; the sum of the heats of fusion and vaporization. 620 calories per g.
Latent heat of vaporization	amount of heat required to change 1 g of liquid to a vapor without a change in temperature. For water the latent heat of vaporization is 540 calories at a temperature of 100°C.
μm	micrometer, a unit of length equal to 1×10^{-6}m (.001 mm); replaces the term (μ) micron in international standards.
μ	micron, previously used as a unit of length equal to .001 mm. Some still use this term.

Molar solution	solution containing 1000 ml of a solvent to 1 mole of solute.
Mole	mass numerically equivalent to its molecular weight, expressed in grams.
Molecular volume	volume occupied by 1 mole of a substance.
Molecular weight	weight of a molecule of a substance, which is the combined atomic weights of that substance.
Molecule	smallest unit of matter that can exist and retain all the properties of the original substance.
pH	negative logarithm of the hydrogen ion concentration. Water at 25°C has an H^+ concentration of 10^{-7} and an OH ion concentration of 10^{-7} moles per l; hence the pH of water at 25°C is 7, neither acid nor base.
Poise	coefficient of viscosity; that tangential force required to maintain measured difference in velocity between two parallel planes separated by 1 cm of fluid. 1 poise = 1 dyne/sec/cm.
Poiseuille's postulation	as used here, the resistance to the flow of a vapor through a tube $Q = \pi a^4 p/8\eta l$ where l = the length of the tube, p = the pressure difference between the ends of the tube, Q = the rate of flow and η = the viscosity of the gas, a = the radius of the tube.
Pressure	force exerted upon a surface stated as a force per unit area, such as psi or dynes per cm^2; also may be measured by a column of mercury. As used here, is stated in mm Hg or micrometers (μm Hg).
Radiation	propagation or transmission of energy through space in a manner that may be graphically represented as wave fronts; here used principally to describe transfer of heat through a vacuum.
Raoults Law	when substances that are nonvolatile and are not electrolytes are dissolved in a given weight of a given solvent, they lower the solvent's freezing point, raise the boiling point, and re-

duce the vapor pressure equally for all such solutes.

Supercooling	cooling below the freezing point without crystallization or solidification.
Surface tension	lattice of molecular attraction that exists on the surface of a liquid or at the interface between two fluids; that force that causes formation of water droplets.
Temperature	measure of heat; manifestation of transitional kinetic energy of the molecules of a substance excited by heat agitation. The Centigrade degree is 1/100th of the difference between the freezing and boiling points of water. Absolute temperature is based upon the Kelvin scale where $0°A$, or $0°K$, is equal to $-273.159°C$.
Triple point	conditions of temperature and pressure in which a solid, liquid, and vapor of a substance are in equilibrium. The triple point for water is $0°C$ at a pressure of 4.579 mm Hg.
Vapor pressure	pressure exerted when the vapor of a substance is in equilibrium with that substance.
Viscosity	characteristic of fluids that offers resistance to change in shape; the property of a fluid that resists internal flow; measured in dynes per cm^2 (see poise).

REFERENCES

Altmann, R. 1890. *Die elementarorganismen und ihre beziehungen zu den zellen.* Leipzig: Veit & Co.

Anderson, T. F. 1954. "Preservation of structure in dried specimens." *Proceedings, third international conference on electron microscopy.* London: Royal Microscopic Society, p. 122.

Andrews, D. H. and Kokes, R. J. 1965. *Fundamental chemistry.* New York: John Wiley & Sons.

Asahina, E. 1966. "Freezing and frost resistance of insects." In *Cryobiology,* edited by H. T. Meryman. New York, London: Academic Press.

Bayliss, W. M. 1915. *Principles of general physiology.* London: Longmans, Green, & Co.

Becquerel, P. 1936. "La vie latente de quelques algues et animaux inferieurs aux basse temperatures et la conservation de la vie dans l'univers." *Comptes, rendus 202,* pp. 978-981.

———. 1938. "La congelation et la synerese." *Comptes, rendus 206,* p. 1587.

———. 1950. "La suspension de la vie des spores des moisissures dessechees dans la vide vers la zero absolu." *Comptes, rendus 231,* pp. 1392-1394.

———. 1951. "La suspension de la vie au confins du zero absolu entre 0.0075 K. et 0.047 K." *Proceedings eighth international congress of refrigeration,* p. 326.

Becket, L. G. 1954. *Biological applications of freezing and drying,* edited by R. J. C. Harris. New York: Academic Press, p. 285.

Bell, L. G. E. 1956. "Freeze-drying." *Physical techniques in biological research,* edited by G. Oster and W. W. Polister. New York: Academic Press, 3:1.

Bensley, R. R. and Gersh I. 1933. (A) "Studies on cell structure by the freeze-drying method. I. Introduction. II. The nature of mitochondria in the hepatic cell of *Amblystomae.*" *Anatomy record* 57:205-237.

Bensley, R. R. 1933. (B) "Studies on cell structure by the freeze-drying method. IV. The structure of the inter-kinetic and resting nuclei." *Anatomy record* 58:1-15.

Bensley, R. R. and Hoerr, N. L. 1934. (C) "Studies on cell structures by the freeze-drying method. V. The chemical basis of the organization of cells." *Anatomy record* 60:251-266.

Bird, K. 1968. "The freeze-drying industry: projections of capital and labor requirements." *U.S. Department of Agriculture Marketing Economics Division Report,* vol. 70.

Blum, M. S. and Woodring, J. P. 1963. "Preservation of insect larvae by vacuum dehydration." *Journal of Kansas entomology society* 36:96-101.

———. 1963. "Freeze-drying of spiders and immature insects." *Annals, entomology society of America* 56:138–141.

Braddick, H. J. J. 1954. *The physics of experimental method.* New York: Prentice-Hall.

Brady, B. L. 1959. *Recent research in freezing and drying,* edited by A. S. Parks and A. U. Smith. Oxford, England: Blackwell, vol. 234.

Bullivant, S. and Ames, A. 1966. "A freeze fracture replication method for electron microscopy." *Journal of cell biology* 29:435–447.

Challis, L. J and Wilks, J. 1958. "Heat transfer between solid and liquid helium II." *Proceedings of Kammerlingh Onnes conference on low temperature physics, physica suppliment deel* 24:145.

Chambers, R. and Hale, H. P. 1932. "The formation of ice in protoplasm." *Proceedings of the Royal Society,* Series B. 110:336–352.

Chemical Rubber Co. 1969. *Handbook of chemistry and physics,* edited by R. C. Weast. Cleveland, Ohio.

Chichester, C. O. 1964. In *Lyophilisation-freeze-drying,* edited by L. Rey. Paris: Herman, p. 573.

Corridon, C. A. 1962. "Freeze-drying of foods." A list of selected references. *Library list no. 77 National Agricultural Library,* U.S. Department of Agriculture.

Daniels, F., Mathews, J. H., and Williams, J. W. 1949. *Experimental physical chemistry,* 4th ed. New York: McGraw Hill Book Co.

Davies, D. A. L. 1954. "On preservation of insects by drying in vacuo at low temperature." *Entomologist* 87:34.

Davies, D. A. L. and Baugh, V. S. G. 1956. "Preservation of animals and plants by drying in frozen state." *Nature* 77:657–758, no. 2.

Davies, D. A. L. 1962. "The preservation of larger fungi." *Freeze-drying transcript of British Mycology Society* 45:424–428.

Decareau, R. V. 1962. *Freeze-drying of foods,* edited by F. R. Fisher. Washington, D.C.: National Academy of Science, vol. 147.

Diamond, L. S., Meryman, H. T., and Kafig, E. 1963. *Culture collections: perspectives and problems.* University of Toronto Press, p. 198.

Dushman, S. and Lafferty, J. M. 1962. *Scientific foundations of vacuum techniques.* New York: John Wiley & Sons.

Engstrom, A. and Finian, J. B. 1958. *Biological ultrastructure.* New York: Academic Press.

Feder, N. and Sidman, R. L. 1958. "Methods and principles of fixation by freeze substitution." *Journal of biophysics and biochemistry,* Cytology, 4:592–600.

Fernandez-Moran, H. 1950. "An electonmicroscope study of rat and frog sciatic nerves." *Experimental cell research* 1:309–340.

———. 1959. 'Cryofixation and supplimentary low temperature preparation techniques applied to the study of tissue ultrastructure." Program for the 17th annual meeting, Electron Microscopy Society of America. *Journal of applied physics* 30:2038, no. 12.

———. 1960. "Low temperature preparation techniques for electron microscopy of biological specimens based on rapid freezing with liquid helium II." *Annals, New York Academy of Science* 85:689–713.

Flink, J. 1971. "Conservation of water-damaged written documents by freeze-drying." *Nature*, vol. 234, no. 5329.

Flosdorf, E. W. and Mudd, S. 1935. "Procedure and apparatus for preservation in 'Lyophile' form of serum and other biological substances." *Journal of immunology* 29:389-425.

Flosdorf, E. W., Stokes, F. J., and Mudd, S. 1940. "Desivac process for drying from frozen state." *Journal of American Medical Association* 115:1095.

Flosdorf, E. W. 1949. *Freeze-drying*. New York: Reinhold Book Co.

Gersh, I. 1932. "The Altmann technique for fixation by drying when freezing." *Anatomy records* 53:1-5.

Gersh, I. and Stephenson, J. L. 1954. "Freezing and drying of tissues for morphological and histochemical studies." *Annals, New York Academy of Science* 85:329-384.

Goldblith, S. A. 1964. In *Lyophilisation-freeze-drying*, edited by L. Rey. Paris: Herman, p. 527.

Goodspeed, T. H. 1934. "Applications of the Altmann technique to plant cytology." *Proceedings of the National Academy of Sciences* 20:495-501.

Goodspeed, T. H. and Uber, F. M. 1935. Ibid II. Character of fixation. University of California, *Publications in botany* 28:2332.

Greaves, R. I. N. 1954. *Biological applications of freezing and drying*, edited by R. J. C. Harris. New York: Academic Press, p. 87.

———. 1960. (a) *Annals, New York Academy of Science* 85:682, (b) 85:723, (c) *Recent research in freezing and drying*, edited by A. S. Parks and A. U. Smith. Oxford, England: Blackwell, vol. 203.

———. 1962. *Progres en lyophilisation*, edited by L. Rey. Paris: Herman, vol. 167.

———. 1964. In *Lyophilisation-freeze-drying*, edited by L. Rey, (a) p. 171, (b) p. 407. Paris: Herman.

Hackenberg, U. 1962. *Progres recents en lyophilisation*, edited by L. Rey. Paris: Herman, vol. 99.

———. 1964. In *Lyophilisation-freeze-drying*, edited by L. Rey. Paris: Herman, p. 99.

Haggis, G. H. 1961. "Electron microscope replicas from the surface of fractured through frozen cells." *Journal of biophysics and biochemistry*, Cytology, 9:841-852.

Hall, C. E. 1950. "A low temperature replica method for electron microscopy." *Journal of applied physics* 21:61-62.

Hanson, S. W. F., editors. 1961. *Accelerated freeze-drying method for food preservation*. London, H.S.M.O.

Harris, R. H. 1964. "Vacuum dehydration and freeze-drying of entire biological specimens." *Annals & magazine, natural history*, Series 13, 7:65-74.

———. 1968. "A new apparatus for freeze-drying whole biological specimens." *Medical biology illustrated* 18:180-182.

Harrow, B. 1951. *Textbook of biochemistry*. Philadelphia: W. E. Saunders Co.

Haskings, R. H. 1960. "Freeze-drying of macro fungi for display." *Mycologia* 50:161–164.

Hoerr, N. L. 1936. "Cytological studies by the Altmann-Gersh freeze-drying method. I. Recent advances in the technique." *Anatomy record* 65:239–317.

Hower, R. O. 1962. "Freeze-drying biological specimens." *Smithsonian Institution information leaflet, No. 324.*

———. 1964. "Freeze-drying biological specimens." *Museum news, technical supplement*, vol. 1, no. 1.

———. 1967. "The freeze-dry preservation of biological specimens." *Proceedings of the U.S. National Museum*, vol. 119, no. 3549.

———. 1969. "Freeze-dry preservation of biological museum specimens." *American taxidermist*, vol. 2, no. 2.

———. 1970. "Advances in freeze-drying." *Curator*, vol. 13, no. 2.

Jacob, S. W., Owen, O. E., Collins, S. C., and Dunphy, J. E. 1958. 'Survival of normal human tissues frozen to $-272.2°$ C." *Transplantation bulletin* 5:248.

Jordan, R. C. and Priester, G. B. 1948. *Refrigeration and air conditioning.* New York: Prentice-Hall.

Keenan, J. H. 1941. *Thermodynamics.* New York: John Wiley & Sons.

Kistler, S. S. 1936. "The measurement of bound water by the freezing method." *Journal of the American Chemical Society* 58:901–907.

Kuprianof, J. 1964. In *Lyophilisation-freeze-drying*, edited by L. Rey. Paris: Herman.

Lambert, J. E. and Marshall, W. R. Jr. 1962. *Freeze-drying of foods*, edited by F. R. Fisher. Washington, D.C.: National Academy of Science.

Love, M. R. 1966. "The freezing of animal tissue." in *Cryobiology*, edited by H. T. Meryman. New York, London: Academic Press.

Luyet, B. J. 1951. "Survival of cells, tissues and organisms after ultra rapid freezing." *Freezing and drying*, edited by R. C. J. Harris. England Institute of Biology, pp. 77–89.

———. 1962. In *Freeze-drying of foods*, edited by F. R. Fisher. Washington, D.C.: National Academy of Science, p. 194.

MacKenzie, A. P. and Luyet, B. J. 1963. "An electron microscope study of the fine structure of very rapidly frozen blood plasm." *Biodynamica* 9:147–164.

———. 1964. "Apparatus for freeze-drying at very low controlled temperatures." (AFBR freeze-drying apparatus model 1), *Biodynamica* 9:178–191.

MacKenzie, A. P. 1965. "Factors affecting the mechanism of transformation of ice into water vapor in the freeze-drying process." *Annals, New York Academy of Science meeting* 125:522–547.

Mann, G. 1902. *Physiological histology methods and theory.* Oxford: Clarendon Press.

Mathews, A. P. 1925. *Physiological chemistry*, 4th ed. New York: W. Wood & Co.

Menz, L. J. and Luyet, B. J. 1961. "An electron microscope study of the distribution of ice in a single muscle fiber frozen rapidly." *Biodynamica* 8:261-294.

Mercie, F. L. 1948. 'Preparation des collections vegetables sous vide." (Nouvelles Observations). *Bulletin, Society of Botany, France* 95:38-43.

Meryman, H. T. 1950. "Replication of cells and liquids by vacuum evaporation." *Journal of applied physics* 21:68.

Meryman, H. T. and Kafig, E. 1955. "The study of frozen specimens ice crystals and ice crystal growth by electron microscopy." Naval Medical Research Institute *Project report NM000-18.01.09* 13:529-544.

Meryman, H. T. and Platt, W. T. 1955. "The distribution and growth of ice in frozen mammalian tissue." Naval Medical Research Institute *Project Report NM000-18.01.08* 12:1-3.

Meryman, H. T. 1960. In *Recent research in freezing and drying*, edited by A. S. Parks, and A. U. Smith. Oxford, England: Blackwell.

————. 1960. "The preparation of biological museum specimens by freeze-drying." *Curator* 1:5-19, no. 3.

————. 1961. *The preparation of biological museum specimens.* Naval Medical Research Institute, Bethesda, Maryland.

————. 1962. In *Freeze-drying of foods*, edited by F. R. Fisher. Washington, D.C.: National Academy of Science, p. 225.

————. 1964. In *Lyophilisation-freeze-drying*, edited by L. Rey. Paris: Herman.

————. 1965. In "Progress in refrigeration science and technology." *Proceedings XI. International congress of refrigeration in Munich.* Oxford Pergamon, p. 1609.

————. 1966. *Cryobiology.* London, New York: Academic Press.

Moor, H. 1964. "Die gefrier-fixation lebender zellen und ihre anwendung in der elektronenmikroskopie." *Zeitschrift zellforsch* 62:546-580.

Muggleton, P. W. 1963. "The preservation of cultures." *Progress in industrial microbiology* 4:189-214.

Muller, H. R. 1957. "Gefriertrocknung als fixierungsmethode an phlanzellen." *Journal, ultrastructure research* 1:109-137.

Parks, A. S. 1951. "Preservation of spermatazoa, red blood cells, and endocrine tissue." *Freezing and drying*, edited by R. C. J. Harris. England Institute of Biology.

Patten, S. F. and Brown, K. A. 1958. "Freeze solvent substitution technique." *England Lab* 7:209.

Rapatz, G. and Luyet, B. J. 1960. "Microscopic observations on the development of the ice stage in the freezing of blood." *Biodynamica* 8:197-239.

Rey, L., editor. 1960. (a) *Traite de lyophilisation.* Paris: Herman. (b) *Annals, New York Academy of Science* 85:510.

————, editors. 1964. *Lyophilisation-freeze-drying.* Paris: Herman.

Rightsel, W. A. and Greiffe, D. 1967. "Freezing and freeze-drying of viruses." *Cryobiology* 3:423.

Romeis, B. 1932. *Taschenbuch der mikroskopischen technik*, 2d ed. Munchen and Berlin: R. Oldenbourg.

Rowe, T. 1960. "The theory and practice of freeze-drying." *Annals, New York Academy of Science* 85:679-681.

———. 1964. In *Lyophilisation-freeze-drying*, edited by L. Rey. Paris: Herman.

Salwin, H. and Slawson, V. S. 1959. "Freeze-drying of foods." *Food Technology* 13:2038.

Schmitt, F. O. 1944. "Tissue ultrastructure analysis." *Medical physics*, edited by Glasser. Chicago: Yearbook, vol. 1586.

Scott, G. H. 1934. "A critical review of the methods of micro incineration." *Protoplasma* 20:133-151.

Scott, R. B. 1959. *Cryogenic engineering*. Princeton: Van Nostrand.

Scott, W. J. 1960. "Recent research in freezing and drying." *Annals, New York Academy of Science* 85:188.

Sherman, J. K. and Kim, K. S. 1967. "Corellation of cellular ultrastructure, before freezing, while frozen, and after thawing, in assessing freeze-thaw induced injury." *Cryobiology* 4:61-74.

Simpson, W. L. 1941. "An experimental analysis of the Altmann technique of freeze-drying." *Anatomy record* 80:173-186.

Singer, J. H. 1954. *Biological applications of freezing and drying*, edited by R. C. J. Harris. New York: Academic Press, p. 151.

Sjostrand, F. S. 1951. "Freeze-drying of tissue for cell analysis by light electron microscopy." *Freezing and drying*, edited by R. C. J. Harris. England Institute of Biology, pp. 177-188.

Sjostrand, F. S. and Baker, R. F. 1958. "Fixation by freeze-drying for electron microscopy of tissue cells." *Journal of ultrastructure research* 1:239-246.

Smith, A. U. 1954. "Effects of low temperatures on living cells and tissues." *Biological applications of freezing and drying*, edited by R. C. J. Harris. New York: Academic Press.

Stadelman, E. J. 1959. "The use of Mercie's method of freeze-drying for the preparation of fungi for demonstration." *Proceedings of IX international congress of botany*, Montreal, 11:376.

Steere, R. L. 1957. "Structural detail by electron microscopy in frozen biological specimens." *Journal of biophysics and biochemistry*, Cytology, 3:46-60.

Stephenson, J. L. 1956. "Ice crystal growth during the rapid freezing of tissues," *Journal of biophysics and biochemistry*, Cytology, 2:45-52, (4).

Strong, J. 1953. *Procedures in experimental physics*. New York: Prentice-Hall.

Sylven, B. 1951. "Effects of freezing and drying compared with common fixation procedures." *Freezing and drying*, edited by R. C. J. Harris. England Institute, Biology.

Food and Container Institute, Technical Services Office. 1960. *Freeze dehydration of foods*. Chicago.

Williams, R. C. 1953. "A method of freeze-drying for electron microscopy." *Experimental cell research* 4:188-201.

————. 1954. "The applications of freeze-drying to electron microscopy." *Biological application of freezing and drying*, edited by R. C. J. Harris. New York: Academic Press, pp. 303–328.

Williams, V. R. and Williams, H. B. 1973. *Basic physical chemistry for the life sciences*. San Francisco: W. H. Freeman & Co.

Wollaston, W. H. 1813. "On a method of freezing at a distance." *Royal Society of Philosophy Transactions*. London, pp. 71–74.

INDEX A
Authors

Abraham J. L., 100
Altmann, R., 19, 20, 183
Ames, A., 184
Anderson, T. F., 183
Andrews, D. H., 25, 183
Asahina, E., 183
Baugh, V. S. G., 184
Baker, R. F., 188
Bayliss, W. M., 19, 183
Bensley, R. R., 19, 183
Bird, K., 183
Blum, M. S., 80, 183
Braddick, H. J. J., 184
Brady, B. L., 184
Brown K. A., 187
Bullivant, S., 184
Challis, L. J., 184
Chambers, R., 184
Chichester, C. O., 184
Collins, S. C., 186
Corridon, C. A., 184
Daniels, F., 184
Davies, D. A. L., 20, 21, 184
Decareau, R. V., 184
Diamond, L. S., 184
Dunphy, J. E., 184
Dushman, S., 184
Engstrom, A., 184
Feder, N., 184
Fernandez-Moran, H., 184
Finian, J. B., 184
Flink, J., 185
Flosdorf, E. W., 185
Gersh, I., 19, 20, 183, 185
Goldblith, S. A., 185
Goodspeed, T. H., 20, 185
Greaves, R. I. N., 185
Greiffe, D., 187
Hackenberg, U., 185
Haggis, G. H., 185
Hale, H. P., 184
Hanson, S. W. F., 185
Harris, R. H., 16, 21, 83, 185
Harrow, B., 23, 185
Haskins, R. H., 186
Hoerr, N. L., 19, 20, 183, 186

Howard, C., 16, 28
Hower, R. O., 13, 21, 28, 186
Jacob, S. W., 186
Jordan, R. C., 186
Keenan, J. H., 186
Kim, K. S., 188
Kistler, S. S., 186
Kokes, R. J., 25, 183
Kuprianof, J., 186
Lafferty, J. M., 184
Lambert, J. E., 186
Love, M. R., 186
Luyet, B. J., 186, 187
MacKenzie, A. P., 186
Mann, G., 19, 186
Mathews, A. P., 186
Menz, L. J., 186
Mercie, F. L., 20, 187
Meryman, H. T., 16, 21, 184, 187
Moor, H., 187
Mudd, S., 185
Muggleton, P. W., 187
Muller, H. R., 187
Owen, O. E., 187
Parks, A. S., 187, 184
Patten, G., 187
Priester, G. B., 186
Rapatz, G., 187
Rey, L., 184, 187
Rightsel, W. A., 187
Romeis, B., 19, 187
Rowe, T., 188
Salwin, H., 188
Schmitt, F. O., 188
Scott, R. B., 20, 188
Sherman, J. K., 188
Sidman, R. L., 184
Simpson, W. L., 20, 188
Singer, J. H., 188
Sjostrand, F. S., 20, 188
Slawson, V. S., 188
Smith, A. U., 184, 185, 188
Stadelmann, E. J., 188
Steer, R. L., 188
Stephenson, J. L., 185, 188
Stokes, F. J., 185

INDEX B
Subject

pressure-temperature relationships
of refrigerants, table, 175
protein denaturization, 88
protection, insect, 107-8
protozoa, 117-18
pump down chart, vacuum, 36
pumpkinseed, 67
pump, rotary oil, 32
pump, rotary piston, 33
pump, vacuum, 31
pupils, heat formed, eyes, 73-74
Rallus limicola limicola, drying
chart, 160
rapid freezing, 23
rattlesnake, 64
red-bellied woodpecker, drying
chart, 157
reduction, freezing point, 125
references to freeze-drying, 183-89
refrigerants, 42-43, 175
refrigerants, azeotropic mixtures, 43
refrigerant liquid-vapor phasing, 42
refrigerant-oil mixture, 45
refrigerants, temperature-pressure
relationship chart, 175
refrigerated coil, 31, 48-53
refrigerated condensing surface, 31,
48-54
refrigerated condenser, 31, 48-54
refrigeration, basic concepts of,
42-43
refrigeration compressor, 42-43
refrigeration compressor, water
cooled, 43
refrigeration expansion valves, 44
refrigeration filter dryer, 45
refrigeration oil, 45
refrigeration sight glass, 45
removing blood stains, 62
reptiles, 63-66
robin, 61
rotary oil pump, 32
rotary piston pump, 33
ruby-throated hummingbird, drying
chart, 159
salts, concentration of, 23
salted beef, 124
scanning electron microscopy,
109-21
schematic of large freeze-dry
system, 51
sealing oil vacuum, 37
sharp-shinned hawk, 61

shipping frozen specimens, 108
sight glass, refrigeration, 45
simple freeze-dry apparatus, 47
slate-colored junco, drying chart,
158
small mammals, 56
specimen chamber, 31, 47, 49-52
spherical model, ice front, 131
Spizella passerina passerina, drying
chart, 161
sponge, 72
Skogsbergiella spinifera (Skogsberg),
115
snakes, 63-64
soldering aid, 58
Spinacopia bisetula Kornicker, 113
spleen, 92
Stagmomantis carolina, 82
stains blood, removing, 62
starling, 62
stomach, human, 86
Strix varia, 60
Strix varia, drying chart, 156
structure of ice, 25
Sturnus vulgaris vulgaris, 62
sublimation, 26
sublimation boundary, 29
sublimation of ice, 26
sublimation, latent heat of, 26
Suctoria 117-18
sulphonamide derivative, (Edolan),
108
support, wire, 55-56
surface area at drying front, 29, 131
surface to mass ratio, 131
surface to mass ratio, chart, 133
surface tension, 23, 26
survey of freeze-dry research, 15
system, freeze-dry, 31-54
tables, list of, 12
tabulated tortoise, 65
Tamias striatus, drying chart, 151
temperature conversion table, 173
temperature-pressure relationships
of refrigerants, chart, 175
tetrahedral structure of ice, 25
tetra methylthiuram disulfide, 108
thermal insulation, 45, 46
thermocouple vacuum gauge, 41
thermodynamic properties of water,
26
thermostatic expansion valves, 44
thickness of dried tissue, 131

thyroid gland, drying chart, 143
T.M.T.D., 108
tortoise, tabulated, 65
transport, ion, 127
transfer of water molecules, 29
trap, vapor, 31, 48–49
triple point, 27
tumors, brain, 106
Turdus migratorius, 61
turtles, 64–65
ultimate vacuum, 32
ultrasonic washing, 111
Underwriters Laboratories
    Classification of Comparative
    Life Hazards of Gasses and
    vapors, 174
vacuum dehydration, 16, 19, 80
vacuum former, 78
vacuum forming eye master, 78
vacuum gauges, 39–41
vacuum gauge, thermocouple, 41
vacuum line dimensions, 37
vacuum pump, 31
vacuum valve, 48, 49, 54
vacuum pump capacity, 35
vacuum pump displacement, 32
vacuum pump down chart, 36
vacuum pump down factor, 36
vacuum sealing oil, 37
vacuum table, 32
valve, heart, 103–104
valve, vacuum, 48, 49, 54
vapor conductance, 38–39
vapor flow, 37
vaporization, 25
vaporization, heat of, 25
vapor line, 37
vapor line conductance, chart, 39
vapor pressure calculation, Hower, 27
vapor pressure calculation, Kelly, 27

vapor pressure, equilibrium, 26
vapor pressure, table, 26–27
vapor pressure, water, 26–27
vapor trap, 31, 48–49
Varanus salvator, 66
Vargula subantarctica Kornicker, 116
variations in freezing during electrolysis, 128
vented exhaust, 34
Verio philadelphicus, drying chart, 163
virginia rail, drying chart, 160
viscous flow, 37
washing apparatus, Hope, 110, 111, 117
washing, ultrasonic, 111
water cooled refrigeration compressor, 43
water in biological tissues, 23
water hydrogen bonding, 25
water, microcrystallites in, 25
water, thermodynamic properties, 25
water vapor pressure, 27
whale louse, 70
Wilsonia citrina, drying chart, 162
wire gauge, table, 56
wire preparation, 55–56
wire support, 55
wood from ship's timber, 127
wooden tool handle marine archeology, 126
xerograph of enlarged heart, 103
yellow breasted chat, 59
yellow-shafted flicker, 60
Zeniadura macroura, drying chart, 155
zoological specimens, collecting of, 58
zoological specimens, preparation of, 55–72, 74, 80–83